SpringerBriefs in Electrical and Computer Engineering

Control, Automation and Robotics

More information about this series at http://www.springer.com/series/10198

Andrew D. Lewis

Tautological Control Systems

Springer

Andrew D. Lewis
Department of Mathematics and Statistics
Queen's University
Kingston
Canada

ISSN 2192-6786 ISSN 2192-6794 (electronic)
ISBN 978-3-319-08637-8 ISBN 978-3-319-08638-5 (eBook)
DOI 10.1007/978-3-319-08638-5

Library of Congress Control Number: 2014942664

Springer Cham Heidelberg New York Dordrecht London

Printed on acid-free paper

Springer is part of Springer Science+Business Media (www.springer.com)

Preface

The lack of feedback-invariance of mathematical formulations of nonlinear control theory has been a thorn in the side of understanding the basic structure of control systems. Moreover, it is a thorn whose presence has largely come to be accepted, and this has prohibited a complete understanding of certain fundamental structural problems for nonlinear systems. One way to understand the issue is as follows: just as an explicit parameterisation of system dynamics by state, i.e., a choice of coordinates, can impede the identification of general structure, so it is too with an explicit parameterisation of system dynamics by control. However, such explicit and fixed parameterisation by control is commonplace in control theory, leading to definitions, methodologies and results that depend in unexpected ways on control parameterisation. This unexpected dependence makes it virtually impossible to comprehensively address the fundamental structural problems in control theory, such as controllability and stabilisability.

In this monograph, we present a framework for modelling systems in geometric control theory in a manner that does not make any choice of parameterisation by control; the systems are called 'Tautological Control Systems'. For the framework to be coherent, it relies in a fundamental way on topologies for spaces of vector fields. As such, we take advantage of recent characterisations of topologies for spaces of vector fields possessing a variety of degrees of regularity: finitely differentiable; Lipschitz; smooth; real analytic. As part of the presentation, therefore, locally convex topologies for spaces of vector fields are comprehensively reviewed. It is these locally convex topologies that provide for the unified treatment of time-varying vector fields that underpin the approach.

This monograph presents simply the foundations of the approach, as well as the basic results that indicate the structural attributes of 'Tautological Control Systems'. In particular, we are able to prove the feedback-invariance of the approach. Future work will involve using this feedback-invariant approach to address the basic problems of control theory, e.g., controllability, stabilisability, and optimality.

The author was a Visiting Professor in the Department of Mathematics at University of Hawaii, Manoa, when the monograph was written, and would like to acknowledge the hospitality of the department, particularly that of Monique Chyba and George Wilkens. The author also thanks his departmental colleague at Queen's, Mike Roth, for numerous useful conversations over the years. While conversations with Mike did not lead directly to results in this monograph, Mike's willingness to chat about complex geometry and to answer ill-informed questions was always appreciated and, ultimately, very helpful. Joint work on topologies for spaces of vector fields with the author's Doctoral student, Saber Jafarpour, was essential to the completion of this work.

Honolulu, May 2014 Andrew D. Lewis

Contents

Series Editors' Biography

Tamer Başar is with the University of Illinois at Urbana-Champaign, where he holds the academic positions of Swanlund Endowed Chair, Center for Advanced Study Professor of Electrical and Computer Engineering, Research Professor at the Coordinated Science Laboratory, and Research Professor at the Information Trust Institute. He received the B.S.E.E. degree from Robert College, Istanbul, and the M.S., M.Phil, and Ph.D. degrees from Yale University. He has published extensively in systems, control, communications, and dynamic games, and has current research interests that address fundamental issues in these areas along with applications such as formation in adversarial environments, network security, resilience in cyber-physical systems, and pricing in networks.

In addition to his editorial involvement with these *Briefs*, Başar is also the Editor-in-Chief of *Automatica*, Editor of two Birkhäuser Series on *Systems & Control* and *Static & Dynamic Game Theory*, the Managing Editor of the *Annals of the International Society of Dynamic Games* (ISDG), and member of editorial and advisory boards of several international journals in control, wireless networks, and applied mathematics. He has received several awards and recognitions over the years, among which are the Medal of Science of Turkey (1993); Bode Lecture Prize (2004) of IEEE CSS; Quazza Medal (2005) of IFAC; Bellman Control Heritage Award (2006) of AACC; and Isaacs Award (2010) of ISDG. He is a member of the US National Academy of Engineering, Fellow of IEEE and IFAC, Council Member of IFAC (2011–14), a past president of CSS, the founding president of ISDG, and president of AACC (2010–11).

Antonio Bicchi is Professor of Automatic Control and Robotics at the University of Pisa. He graduated at the University of Bologna in 1988 and was a postdoc scholar at M.I.T. A.I. Lab between 1988 and 1990.
His main research interests are in:

- dynamics, kinematics, and control of complex mechanical systems, including robots, autonomous vehicles, and automotive systems;
- haptics and dextrous manipulation; and

- theory and control of nonlinear systems, in particular hybrid (logic/dynamic, symbol/signal) systems.

He has published more than 300 papers on international journals, books, and refereed conferences.

Professor Bicchi currently serves as the Director of the Interdepartmental Research Center "E. Piaggio" of the University of Pisa, and President of the Italian Association or Researchers in Automatic Control. He has served as Editor in Chief of the Conference Editorial Board for the IEEE Robotics and Automation Society (RAS), and as Vice President of IEEE RAS, Distinguished Lecturer, and Editor for several scientific journals including the *International Journal of Robotics Research*, the *IEEE Transactions on Robotics and Automation*, and *IEEE RAS Magazine*. He has organized and co-chaired the first WorldHaptics Conference (2005), and Hybrid Systems: Computation and Control (2007). He is the recipient of several best paper awards at various conferences, and of an Advanced Grant from the European Research Council. Antonio Bicchi has been an IEEE Fellow since 2005.

Miroslav Krstic holds the Daniel L. Alspach chair and is the founding director of the Cymer Center for Control Systems and Dynamics at University of California, San Diego. He is a recipient of the PECASE, NSF Career, and ONR Young Investigator Awards, as well as the Axelby and Schuck Paper Prizes. Professor Krstic was the first recipient of the UCSD Research Award in the area of engineering and has held the Russell Severance Springer Distinguished Visiting Professorship at UC Berkeley and the Harold W. Sorenson Distinguished Professorship at UCSD. He is a Fellow of IEEE and IFAC. Professor Krstic serves as Senior Editor for *Automatica and IEEE Transactions on Automatic Control* and as Editor for the Springer series *Communications and Control Engineering*. He has served as Vice President for Technical Activities of the IEEE Control Systems Society. Krstic has co-authored eight books on adaptive, nonlinear, and stochastic control, extremum seeking, control of PDE systems including turbulent flows, and control of delay systems.

Chapter 1
Introduction, Motivation, and Background

One can study nonlinear control theory from the point of view of applications, or from a more fundamental point of view, where system structure is a key element. From the practical point of view, questions that arise are often of the form, "How can we…", for example, "How can we steer a system from point A to point B?" or, "How can we stabilise this unstable equilibrium point?" or, "How can we manoeuvre this vehicle in the most efficient manner?" From a fundamental point of view, the problems are often of a more existential nature, with, "How can we" replaced with, "Can we". These existential questions are often very difficult to answer in any sort of generality.

As one thinks about these fundamental existential questions and looks into the quite extensive existing literature, one comes to understand that the question, "What is a control system?" is one whose answer must be decided upon with some care. One also begins to understand that structure coming from common physical models can be an impediment to general understanding. For example, in a real physical model, states are typically physical quantities of interest, e.g., position, current, quantity of reactant X, and so the explicit labelling of these is natural. This labelling amounts to a specific choice of coordinates, and it is now well understood that such specific choices of coordinate obfuscate structure, and so are to be avoided in any general treatment. In like manner, in a real physical model, controls are likely to have meaning that one would like to keep track of, e.g., force, voltage, flow. The maintenance of these labels in a model provides a specific parameterisation of the inputs to the system, completely akin to providing a specific coordinate parameterisation for states. However, while specific coordinate parameterisations have come (by many) to be understood as a bad idea in a general treatment, this is not the case for specific control parameterisations; models with fixed control parameterisation are commonplace in control theory. In contrast to the situation with dependence of *state* on parameterisation, the problem of eliminating dependence of *control* on parameterisation is not straightforward. In our discussion below we shall overview some of the common models for control systems, and some ways within these modelling frameworks for overcoming the problem of dependence on control parameterisation. As we shall see, the common models all have some disadvantage or other that must be confronted

A.D. Lewis, *Tautological Control Systems*, SpringerBriefs in Control, Automation and Robotics, DOI: 10.1007/978-3-319-08638-5_1

when using them. In this monograph we provide a means for eliminating explicit parameterisation of controls that, we believe, overcomes the problems with existing techniques. Our idea has some of its origins in the work on "chronological calculus" of Agrachev and Gamkrelidze [2] (see also [3]), but the approach we describe here is more general (in ways that we will describe below) and more fully developed as concerns its relationship to control theory (chronological calculus is primarily a device for understanding time-varying vector fields and flows). There are some ideas similar to ours in the approach of Sussmann [64], but there are also some important differences, e.g., our families of vector fields are time-invariant (corresponding to vector fields with frozen control values) while Sussmann considers families of time-varying vector fields (corresponding to selecting an open-loop control). Also, the work of Sussmann does not touch on real analytic systems.

We are interested in models described by ordinary differential equations whose states are in a finite-dimensional manifold. Even within this quite narrow class of control systems, there is a lot of room to vary the models one might consider. Let us now give a brief outline of the sorts of models and methodologies of this type that are commonly present in the literature.

1.1 Models for Geometric Control Systems: Pros and Cons

By this time, it is well-understood that the language of systems such as we are considering should be founded in differential geometry and vector fields on manifolds [3, 7, 11, 30, 36, 50]. This general principle can go in many directions, so let us discuss a few of these. Our presentation here is quite vague and not very careful. In the main body of the work, we will be less vague and more careful.

1.1.1 Family of Vector Field Models

Given that manifolds and vector fields are important, a first idea of what might comprise a control system is that it is a family of vector fields. For these models, trajectories are concatenations of integral curves of vector fields from the family. This is the model used in the development of the theory of accessibility of Sussmann and Jurdjevic [66] and in the early work of Sussmann [61] on local controllability. The work of Hermann and Krener [27], while taking place in the setting of systems parameterised by control (such as we shall discuss in Sect. 1.1.2), uses the machinery of families of vector fields to study controllability and observability of nonlinear systems. Indeed, a good deal of the early work in control theory is developed in this sort of framework, and it is more or less sufficient when dealing with questions where piecewise constant controls are ample enough to handle the problems of interest. The theory is also highly satisfying in that it is very differential geometric, and the work utilising this approach is often characterised by a certain elegance.

However, the approach does have the drawback of not handling well some of the more important problems of control theory, such as feedback (where controls are specified as functions of state) and optimal control (where piecewise constant controls are often not a sufficiently rich class, cf. [18]).

It is worth mentioning at this early stage in our presentation that one of the ingredients of our approach is a sort of fusion of the "family of vector fields" approach with the more common control parameterisation approach to whose description we now turn.

1.1.2 Models with Control as a Parameter

Given the limitations of the "family of vector fields" models for physical applications and also for a theory where merely measurable controls are needed, one feels as if one has to have the control as a parameter in the model, a parameter that one can vary in a quite general manner. These sorts of models are typically described by differential equations of the form

$$\dot{x}(t) = F(x(t), u(t)),$$

where $t \mapsto u(t)$ is the control and $t \mapsto x(t)$ is a corresponding trajectory. For us, the trajectory is a curve on a differentiable manifold M, but there can be some freedom in attributing properties to the control set $\mathcal{C}$ in which u takes its values, and on the properties of the system dynamics F. (In Sect. 3.3 we describe classes of such models in differential geometric terms.) This sort of model is virtually synonymous with "nonlinear control system" in the existing control literature. A common class of systems that are studied are control-affine systems, where

$$F(x, \mathbf{u}) = f_0(x) + \sum_{a=1}^{k} u^a f_a(x),$$

for vector fields $f_0, f_1, \ldots, f_k$ on M, and where the control $\mathbf{u}$ takes values in a subset of $\mathbb{R}^k$. For control-affine systems, there is an extensively developed theory of controllability based on free Lie algebras [6, 40–42, 62, 63]. We will see in Sect. 3.3.2 that control-affine systems fit into our framework in a particularly satisfying way.

The above general model, and in particular the control-affine special case, are all examples where there is an explicit parameterisation of the control set, i.e., the control u lives in a particular set and the dynamics F is determined to depend on u in some particular way. It could certainly be the case, for instance, that one could have two different systems

$$\dot{x}(t) = F_1(x(t), u_1(t)), \quad \dot{x}(t) = F_2(x(t), u_2(t))$$

with exactly the same trajectories. This has led to an understanding that one should study equivalence classes of systems. A little precisely, if one has two systems

$$\dot{x}_1(t) = F_1(x_1(t), u_1(t)), \quad \dot{x}_2(t) = F_2(x_2(t), u_2(t)),$$

with $x_a(t) \in \mathsf{M}_a$ and $u_a(t) \in \mathcal{C}_a$, $a \in \{1, 2\}$, then there may exist a diffeomorphism $\Phi \colon \mathsf{M}_1 \to \mathsf{M}_2$ and a mapping $\kappa \colon \mathsf{M}_1 \times \mathcal{C}_1 \to \mathcal{C}_2$ (with some sort of regularity that we will not bother to mention) such that

1. $T_{x_1}\Phi \circ F_1(x_1, u_1) = F_2(\Phi(x_1), \kappa(x_1, u_1))$ and
2. the trajectories $t \mapsto x_1(t)$ for the first system are in 1–1 correspondence with those of the second system by $t \mapsto \Phi \circ x_1(t)$.[1]

Let us say a few words about this sort of "feedback equivalence". One can imagine it being useful in at least two ways.

1. First of all, one might use it as a kind of "acid test" on the viability of a control theoretic construction. That is, a control theoretic construction should make sense, not just for a system, but for the equivalence class of that system. This is somewhat akin to asking that constructions in differential geometry should be independent of coordinates. Indeed, in older presentations of differential geometry, this was often how constructions were defined: they were given in coordinates, and then demonstrated to behave properly under changes of coordinate. We shall illustrate in Examples 1.1 and 1.2 below that many common constructions in control theory do not pass the "acid test" for viability as feedback-invariant constructions.
2. Feedback equivalence is also a device for classifying control systems, the prototypical example being "feedback linearisation", the determination of those systems that are linear systems in disguise [34]. In differential geometry, this is akin to the classification of geometric structures on manifolds, e.g., Riemannian, symplectic, etc.

In Sect. 5.6 we shall consider a natural notion of equivalence for systems of the sort we are introducing in this work, and we will show that "feedback transformations" are vacuous in that they amount to being described by mappings between manifolds. This is good news, since the whole point of our framework is to eliminate control parameterisation from the picture and so eliminate the need for considering the effects of varying this parameterisation, cf. "coordinate-free" versus "coordinate-independent" in differential geometry. Thus the first of the preceding uses of feedback transformations simply does not come up for us: our framework is naturally feedback-invariant. The second use of feedback transformations, as will be seen in Sect. 5.6, amounts to the classification of families of vector fields under push-forward by diffeomorphisms. This is generally a completely hopeless undertaking, so we will have nothing to say about this. Studying this under severe restrictions using, for example, (1) the Cartan method of equivalence, e.g., [10, 19, 25, 26], (2) the method

[1] We understand that there are many ways of formulating system equivalence. But here we are content to be, not only vague, but far from comprehensive.

of generalised transformations, e.g., [37, 38], (3) the study of singularities of vector fields and distributions, e.g., [34, 51], one might expect that some results are possible.

There is an important facet of our modelling framework that comes out in the preceding paragraph, and to which we wish to draw special attention:

> Our framework is not one for determining *feedback invariants* of control systems. It is one for providing models that have the attribute of *feedback-invariance*.

Thus, while it is tempting to regard what we do as being in the spirit of the work mentioned above on feedback equivalence, the approach and objectives are really quite different.

Let us consider an example that shows how a classical control-theoretic construction, linearisation, is not invariant under even the very weak notion of equivalence where equivalent systems are those with the same trajectories.

Example 1.1 (Linearisation is not well-defined). We consider the two control-affine systems

$$\begin{aligned} \dot{x}_1(t) &= x_2(t), & \dot{x}_1(t) &= x_2(t), \\ \dot{x}_2(t) &= x_3(t)u_1(t), & \dot{x}_2(t) &= x_3(t) + x_3(t)u_1(t), \\ \dot{x}_3(t) &= u_2(t), & \dot{x}_3(t) &= u_2(t), \end{aligned}$$

with $(x_1, x_2, x_3) \in \mathbb{R}^3$ and $(u_1, u_2) \in \mathbb{R}^2$. One can readily verify that these two systems have the same trajectories. If we linearise these two systems about the equilibrium point at $(0, 0, 0)$—in the usual sense of taking Jacobians with respect to state and control [30, p. 172], [43, Sect. 12.2], [50, Proposition 3.3], [56, p. 236], and [59, Definition 2.7.14]—then we get the two linear systems

$$\mathbf{A}_1 = \begin{bmatrix} 0 & 1 & 0 \\ 0 & 0 & 0 \\ 0 & 0 & 0 \end{bmatrix}, \quad \mathbf{B}_1 = \begin{bmatrix} 0 & 0 \\ 0 & 0 \\ 0 & 1 \end{bmatrix}, \quad \mathbf{A}_2 = \begin{bmatrix} 0 & 1 & 0 \\ 0 & 0 & 1 \\ 0 & 0 & 0 \end{bmatrix}, \quad \mathbf{B}_2 = \begin{bmatrix} 0 & 0 \\ 0 & 0 \\ 0 & 1 \end{bmatrix},$$

respectively. The linearisation on the left is not controllable, while that on the right is.

The example suggests that (1) classical linearisation is not independent of parameterisation of controls and/or (2) the classical notion of linear controllability is not independent of parameterisation of controls. Both things, in fact, are true: neither classical linearisation nor the classical linear controllability test are feedback-invariant. Said differently, linearisation and linear controllability, when applied to representatives of an equivalence class of systems, do not allow conclusions that apply to all members of the same equivalence class. This may come as a surprise to some. ∘

This example has been particularly chosen to provide probably the simplest illustration of the phenomenon of lack of feedback-invariance of common control theoretic constructions. Therefore, it should not be a surprise that an astute reader will notice that linearising the "uncontrollable" system about the control $(1, 0)$ rather than the control $(0, 0)$ will square things away as concerns the discrepancy between the two linearisations. But after doing this, the questions of, "What are the proper *definitions* of linearisation and linear controllability?" still remain. Moreover, one might expect that as one moves to constructions in control theory more sophisticated than mere linearisation, the dependence of these constructions on the parameterisation of controls becomes more pronounced. Let us illustrate this with another example.

Example 1.2 (Obstructions to controllability are not feedback-invariant). We again consider two control-affine systems on $\mathbb{R}^3$ with two inputs:

$$\begin{aligned} \dot{x}_1(t) &= u_1(t), & \dot{x}_1(t) &= \tfrac{1}{2}(u_1(t)+u_2(t)), \\ \dot{x}_2(t) &= u_2(t), & \dot{x}_2(t) &= \tfrac{1}{2}(u_1(t)-u_2(t)), \\ \dot{x}_3(t) &= x_1(t)^2 - x_2(t)^2, & \dot{x}_3(t) &= x_1(t)^2 - x_2(t)^2. \end{aligned}$$

Thus both systems have the drift vector field $f_0 = (x_1^2 + x_2^2)\frac{\partial}{\partial x_3}$, the system on the left has control vector fields $f_1 = \frac{\partial}{\partial x_1}$ and $f_2 = \frac{\partial}{\partial x_2}$, and the system on the right has control vector fields $f_1' = \frac{1}{2}(\frac{\partial}{\partial x_1} + \frac{\partial}{\partial x_2})$ and $f_2' = \frac{1}{2}(\frac{\partial}{\partial x_1} - \frac{\partial}{\partial x_2})$. It is pretty obvious that these two systems have the same trajectories. We will apply the well-known "bad bracket test" of Sussmann [63, Theorem 7.3] to study the local controllability of these two control-affine systems. This test says that the system is locally controllable from x_0 if all "bad brackets", i.e., those having an odd number of drift terms and an even number of each of the control vector fields, are a finite linear combination at x_0 of "good brackets" ("good" meaning "not bad") of lower degree, where degree means the total number of terms in the bracket.[2] For the systems at hand, the relevant bad brackets are

$$[f_1,[f_1,f_0]] = 2\frac{\partial}{\partial x_3}, \quad [f_2,[f_2,f_0]] = -2\frac{\partial}{\partial x_3},$$

and

$$[f_1',[f_1',f_0]] = 0, \quad [f_2',[f_2',f_0]] = 0.$$

In each case, the good brackets of lower degree vanish at (0, 0, 0). Thus, for the system on the left, the bad bracket test is inconclusive about local controllability from (0, 0, 0) since its hypotheses do not hold. However, for the system on the right, the hypotheses of the bad bracket test hold (the bad brackets are zero), and so we can conclude that the system is controllable from (0, 0, 0).

Just as Example 1.1 suggests that the classical constructions of linearisation and linear controllability are not feedback-invariant, the preceding calculations show that one of the well-known tests for local controllability is not feedback-invariant. This is well-known in this case, e.g., it has been pointed out (more comprehensively than here) in [11, Example 7.22]. ○

---

[2] Sussmann actually has a more sophisticated notion of degree, but for this example it boils down to the one we give.

As the preceding examples illustrate, the likelihood that a sophisticated construction, made using a specific control parameterisation, is feedback-invariant is quite small, and in any case would need proof to verify that it is. Such verification is not typically part of the standard development of methodologies in control theory. There are at least three reasons for this: (1) the importance of feedback-invariance is not universally recognised; (2) such verifications are generally extremely difficult, nearly impossible, in fact; (3) most methodologies will fail the verification, so it is hardly flattering to one's methodology to point this out. Some discussion of this is made by Lewis [46].

But the bottom line is that our framework simply eliminates the need for any of this sort of verification. As long as one remains within the framework, feedback-invariance is guaranteed. One of the central goals of the monograph is to provide the means by which one does not have to leave the framework to get things done. As we shall see, certain technical difficulties have to be overcome to achieve this, and our work relies crucially on nontrivial recent work of Jafarpour and Lewis [33] on topologies for spaces of vector fields.

1.1.3 Fibred Manifold Models

As we have tried to make clear in the discussion just preceding, the standard model for control theory has the unpleasant attribute of depending on parameterisation of controls. A natural idea to overcome this unwanted dependence is to do with controls as one does with states: regard them as taking values in a differentiable manifold. Moreover, the manner in which control enters the model should also be handled in an intrinsic manner. This leads to the "fibred manifold" picture of a control system which, as far as we can tell, originated in the papers of Brockett [9] and Willems [69], and was further developed by Nijmeier and van der Schaft [48, 49, 68]. This idea has been picked up on by many researchers in geometric control theory, and we point to the papers [5, 12, 16, 45] as illustrative examples.

The basic idea is this. A control system is modelled by a fibred manifold $\pi\colon \mathsf{C} \to \mathsf{M}$ and a bundle map $F\colon \mathsf{C} \to \mathsf{TM}$ over id_{M}:

$$\begin{array}{ccc} \mathsf{C} & \xrightarrow{F} & \mathsf{TM} \\ & \searrow^{\pi} & \downarrow{\scriptstyle \pi_{\mathsf{TM}}} \\ & & \mathsf{M} \end{array}$$

One says that F is "a vector field over the bundle map π". Trajectories are then curves $t \mapsto x(t)$ in M satisfying $\dot{x}(t) = F(u(t))$ for some $t \mapsto u(t)$ satisfying $x(t) = \pi \circ u(t)$. When it is applicable, this is an elegant and profitable model for control theory. For example, for control models that arise in problems of differential geometry or the calculus of variations, this can be a useful model.

The difficulty with the model is that it is not always applicable, especially in physical system models. The problem that arises is the strong regularity of the control set and, implicitly, the controls: C is a manifold so it is naturally the codomain for smooth curves. In practice, control sets in physical models are seldom manifolds, as bounds on controls lead to boundaries of the control set. Moreover, the boundary sets are seldom smooth. Also, as we have mentioned above, controls cannot be restricted to be smooth or piecewise smooth; natural classes of controls are typically merely measurable. These matters become vital in optimal control theory where bounds on control sets lead to bang-bang extremals. When these considerations are overlaid on the fibred manifold picture, it becomes considerably less appealing and indeed problematic. One might try to patch up the model by generalising the structure, but at some point it ceases to be worthwhile; the framework is simply not well suited to certain problems of control theory. Moreover, since the framework includes the standard approach of Sect. 1.1.2 (at least when the control set is a manifold), it shares with the standard approach the lack of feedback-invariance.

1.1.4 Differential Inclusion Models

Another way to eliminate the control dependence seen in the models with fixed control parameterisation is to instead work with differential inclusions. A differential inclusion, roughly (we will be precise about differential inclusions in Sect. 3.3.3), assigns to each $x \in \mathsf{M}$ a subset $\mathscr{X}(x) \subseteq \mathsf{T}_x\mathsf{M}$, and trajectories are curves $t \mapsto x(t)$ satisfying $\dot{x}(t) \in \mathscr{X}(x(t))$. There is a well-developed theory for differential inclusions, and we refer to the literature for what is known, e.g., [4, 17, 58]. There are many appealing aspects to differential inclusions as far as our objectives here are concerned. In particular, differential inclusions do away with the explicit parameterisations of the admissible tangent vectors at a state $x \in \mathsf{M}$ by simply prescribing this set of admissible tangent vectors with no additional structure. Moreover, differential inclusions generalise the control-parameterised systems described in Sect. 1.1.2. Indeed, given such a control-parameterised system with dynamics F, we associate the differential inclusion

$$\mathscr{X}_F(x) = \{F(x, u) \mid u \in \mathcal{C}\}.$$

The trouble with differential inclusions is that their theory is quite difficult to understand if one just starts with differential inclusions coming "out of the blue". Indeed, it is immediately clear that one needs some sort of conditions on a differential inclusion to ensure that trajectories exist. Such conditions normally come in the form of some combination of compactness, convexity, and semicontinuity. However, the differential inclusions that arise in control theory are *highly* structured; certainly they are more regular than merely semicontinuous and they automatically possess many trajectories. Moreover, it is not clear how to develop an *independent* theory of differential inclusions, i.e., one not making reference to standard models for control theory, that captures the desired structure (in Example 5.5–4 we suggest a natural way of characterising a class of differential inclusions useful in geometric control theory). Also, differential inclusions do not themselves, i.e., without additional structure, capture the notion of a flow that is often helpful in the standard control-parameterised models, e.g., in the Maximum Principle of optimal control theory, cf. [65]. However, differential inclusions *are* a useful tool for studying trajectories, and we include them in the development of our new framework in Chap. 5.

1.1.5 The "Behavioural" Approach

Starting with a series of papers [70–72] and the often cited review [73], Willems provides a framework for studying system theory, with an emphasis on linear systems. A comprehensive overview of the behavioural approach to physical system modelling is given in [74]. The idea in this approach is to provide a framework for modelling dynamical systems arising from physical models as subsets of general functions of generalised time taking values in a set. The framework is also intended to provide a mathematical notion of interconnection as relations in a set. In this framework, the most general formulation is quite featureless, i.e., maps between sets and relations in sets. With this level of generality, the basic questions have a computer science flavour to them, in terms of formal languages. When one comes to making things more concrete, say by making the time-domain an interval in $\mathbb{R}$ for continuous-time systems, one ends up with differential-algebraic equations describing the behaviours and relations. For the most part, these ideas seem to have been only reasonably fully developed for linear models [53]; we are not aware of substantial work on nonlinear and/or geometric models in the behavioural approach. It is also the case that the considerations of feedback-invariance, such as we discuss above, are not a part of the current landscape in behavioural models, although this is possible within the context of linear systems, cf. the beautiful book of Wonham [75].

Thus, while there are some idealogical similarities with our objectives and those of the behavioural approach, our thinking in this work is in a quite specific and complementary direction to the existing work on the behavioural point of view.

1.2 An Introduction to Tautological Control Systems

We will, of course, subsequently provide a comprehensive description of tautological control systems and their basic properties. However, it is useful to, at this early stage, give an outline of the approach so that the reader may have a broad idea in mind of what to expect as she reads the more detailed and difficult rigorous presentation.

Let us begin by providing a list of the objectives of our framework, some of which we have described in some detail above, and some of which we have not yet mentioned.

1.2.1 Attributes of a Modelling Framework for Geometric Control Systems

The preceding sections are meant to illustrate some standard frameworks for modelling control systems and the motivation for consideration of these, as well as pointing out their limitations. If one is going to propose a modelling framework, it is important to understand *a priori* just what it is that one hopes to be able to do in this framework. Here is a list of possible criteria, criteria that we propose to satisfy in our framework.

1. Models should provide for control parameterisation-independent constructions as discussed above.
2. We believe that being able to easily handle real analytic systems is essential to a useful theory. In practice, any smooth control system is also real analytic, and one wants to be able to make use of real analyticity to both strengthen conclusions, e.g., the real analytic version of Frobenius's Theorem [47], and to weaken hypotheses, e.g., the infinitesimal characterisation of invariant distributions e.g., [3, Lemma 5.2].
3. The framework should be able to handle regularity in an internally consistent manner. This means, for example, that the conclusions should be consistent with hypotheses, e.g., smooth hypotheses with continuous conclusions suggest that the framework may not be perfectly natural or perfectly well-developed. The pursuit of this internal consistency in the real analytic case requires deep and recent results concerning topologies for spaces of real analytic vector fields [33].
4. The modelling framework should seamlessly deal with distinctions between local and global. Many notions in control theory are highly localised, e.g., local controllability of real analytic control systems. A satisfactory framework should include a systematic way of dealing with constructions in control theory that are of an inherently local nature. Moreover, the framework should allow a systematic means of understanding the passage from local to global in cases where this is possible and/or interesting. As we shall see, there are some simple instances of these phenomena that can easily go unnoticed if one is not looking for them.
5. Our interest is in *geometric* control theory, as we believe this is the right framework for studying nonlinear systems in general. A proper framework for geometric control theory should make it natural to use the tools of differential geometry.
6. While (we believe that) differential geometric methods are essential in nonlinear control theory, the quest for geometric elegance should not be carried out at the expense of a useful *control* theory.

1.2.2 The "Essentials" of Tautological Control Theory

Underneath the technicalities of tautological control systems are two mostly simple ideas.

1.2.2.1 Extracting the Essence of a Control System

In this section we essentially establish a dictionary from "ordinary" control systems to tautological control systems. This dictionary, like an actual dictionary, is not precise. It will be made more precise during the course of the monograph.

Let us suppose that we have a control system $\Sigma = (\mathsf{M}, F, \mathcal{C})$, meaning that M is a manifold of some prescribed regularity (say, smooth or real analytic), $\mathcal{C}$ is the control set, and F describes the dynamics, in the sense that trajectories are absolutely continuous curves $t \mapsto \xi(t) \in \mathsf{M}$ satisfying

$$\xi'(t) = F(\xi(t), \mu(t))$$

for some control $t \mapsto \mu(t) \in \mathcal{C}$. Of course, there needs to be some technical conditions on F to ensure that trajectories exist for reasonable controls. As a minimum, one needs something like: (1) the vector field F^u defined by $F^u(x) = F(x, u)$ is at least continuously differentiable, and maybe smooth or real analytic; (2) $\mathcal{C}$ should be a topological space, and the derivatives of F with respect to x should be jointly continuous functions of x and u. Issues such as this are discussed in the setting of locally convex topologies in the paper [31], and we review this in Sect. 3.3 below. In any case, there are two elements of this model that we wish to pull out. First of all, Σ defines a parameterised set of vector fields

$$\mathscr{F}_\Sigma = \{F^u \mid u \in \mathcal{C}\}.$$

Second, to define a trajectory for Σ, one first specifies an open-loop control $t \mapsto \mu(t)$. This open-loop control then defines a time-varying vector field F^μ by $(t, x) \mapsto F(x, \mu(t))$. A trajectory is then an integral curve of this time-varying vector field. If, for example, F satisfies the conditions above and the control μ is locally essentially bounded, i.e., takes values in a compact subset of $\mathcal{C}$ on compact subsets of time, then the vector field F^μ will satisfy the hypotheses of the Carathéodory existence and uniqueness theorem for integral curves [59, Theorem 54].

Now, using only these two elements of "ordinary" control systems, let us see if we can fashion a methodology for eliminating the parameterisation by the control set $\mathcal{C}$.

First of all, rather than working with the parameterised family of vector fields $\mathscr{F}_\Sigma$, we instead work simply with a subset of vector fields, denoted by $\mathscr{F}$. We may ask that these vector fields have a prescribed regularity, and in this work we allow for smooth, Lipschitz, finitely differentiable, and real analytic dependence on state. This is the easy part. The difficult part is mimicking the effects of specifying an open-loop control $t \mapsto \mu(t)$. As we saw above, this defines a time-varying vector field F^μ. In the case where one simply has a family of vector fields $\mathscr{F}$, one needs to specify a curve $t \mapsto X_t \in \mathscr{F}$ in the family of vector fields. The technical issues that arise are that one needs to be able to describe properties of this curve that ensure that the resulting time-varying vector field $(t, x) \mapsto X_t(x)$ is nice enough to possess integral curves. It turns out that the way to do this is to ask that the curve $t \mapsto X_t$

be "measurable" and "integrable" in an appropriate topology on the space of vector fields. This is carried out in Sect. 3.1 below.

1.2.2.2 The "Presheaf of Vector Fields" Point of View

From the preceding section, we see that subsets of vector fields will feature prominently in our framework. In our work we break with the common approach by talking only about families of vector fields assigned locally. Somewhat precisely (we will remove the "somewhat" in Chap. 4), to each open subset $\mathcal{U} \subseteq \mathsf{M}$ we assign a subset $\mathscr{F}(\mathcal{U})$ of vector fields, and we require that, if open sets $\mathcal{U}$ and $\mathcal{V}$ satisfy $\mathcal{V} \subseteq \mathcal{U}$, then the restrictions of vector fields from $\mathscr{F}(\mathcal{U})$ to $\mathcal{V}$ are members of $\mathscr{F}(\mathcal{V})$. Such a construction is called a "presheaf of sets of vector fields".

The rationale for making constructions such as this may be initially difficult to grasp. Here we point out three reasons for using this structure.

1. Sometimes control theoretic constructions are easily made locally, but global analogues are not so easily understood. Here is an example. A smooth or real analytic distribution is a subset $\mathsf{D} \subseteq \mathsf{TM}$ such that, for each $x \in \mathsf{M}$, there exists a neighbourhood $\mathcal{N}_x$ of x and a family $\mathscr{X}_x$ of smooth or real analytic vector fields on $\mathcal{N}_x$ such that

$$\mathsf{D}_y \triangleq \mathsf{D} \cap \mathsf{T}_y\mathsf{M} = \operatorname{span}_{\mathbb{R}}(X(y) |\ X \in \mathscr{X}_x).$$

The existence, locally, of plenty of vector fields taking values in the distribution follows from the definition. However, the question of whether there are many smooth or real analytic vector fields X on M for which $X(x) \in \mathsf{D}_x$ is not so trivial. In the smooth case, such vector fields can be constructed using cutoff functions. However, in the real analytic case, the existence of globally defined vector fields only follows from nontrivial sheaf theoretic constructions, including, but not limited to, a deployment of Cartan's Theorem A [13]. We shall sketch how this is carried out in Lemma 7.2 when we discuss a tautological control system formulation of sub-Riemannian geometry.
2. The presheaf point of view is the natural one for defining germs. It is to be imagined that many (all?) important local properties of a real analytic system about $x \in \mathsf{M}$ are contained in the germ of the system at x. This is a statement that will not come as a surprise. However, adopting the presheaf formalism makes consideration of such matters an integral part of the framework; indeed, it *forces* one to think carefully about matters of locality. We will see that these matters arise naturally in Chap. 6 when we extend our quite natural notion of trajectory from Sect. 5.3 to one that is actually more natural, but also more difficult to understand. This class of trajectories relies for their definition on the so-called étalé topology of a sheaf, which itself is connected to germs of local sections of the sheaf. In Sect. 7.3 we suggest some places where we anticipate that such considerations

may well contribute to a tautological control system formulation of problems in controllability theory.
3. The third reason for adopting the presheaf formalism is that it aids in addressing questions that come up even in routine control theory. An example of this is seen at the end of Sect. 3.3.3.

The point of the preceding discussion is this: while we expect that many readers will doubt the value of the presheaf formalism that is a part of the tautological control system framework, it is nonetheless the case that this formalism is not hollow in control theoretic terms.

1.3 An Outline of the Monograph

Let us discuss briefly the contents of the monograph.

As mentioned in our requirement 3 in Sect. 1.2.1, we wish to develop a theory that handles regularity in an intelligent manner. This is not as straightforward as it seems. First of all this has simply not been routinely done in control theory in a systematic manner for *any* degree of regularity [32, §VI]. In most any work, the joint dependence of a system on time (for dynamical systems) or control (for control systems) is presented in an incoherent way. The manner in which this is sometimes done is pointed out in clear way in Sects. 3.1.6 and 3.2.6, but the fact of the matter is that the disarray of the manner in which regularity is handled in the literature prohibits any organised critique. In [31, 33] it is revealed that the right way to achieve consistency with respect to regularity is to make use of locally convex topologies for spaces of vector fields. Also, the new class of systems we present in the monograph, i.e., "tautological control systems", are defined in such a way that these locally convex topologies play an essential rôle. We very rapidly overview these topologies in Chap. 2. A reader wishing to understand this material deeply will want to consult [31, 33] and the references cited therein.

In Chap. 3 we present classes of time-varying vector fields and control systems that are essential for our presentation, following the work of Jafarpour and Lewis [31, 33]. We begin in Sect. 3.1 by defining classes of time-varying vector fields and indicating the relationship of these classes to the locally convex topologies of Chap. 2. While in this work we are presenting a framework that we believe should replace the existing framework for nonlinear control theory (at least for the investigation of problems of structure), we do need to establish the relationships between our approach and standard approaches. Thus, in Sect. 3.3 we carefully define classes of "ordinary" control systems that mesh very well with our new class of tautological control systems. We also talk about differential inclusions, as there are useful comparisons to be made between these and our systems.

As we have mentioned a few times, presheaves and sheaves are an essential part of our approach, and in Chap. 4 we give the requisite background. Much of the material that we use from this chapter is quite elementary in character, but, in the constructions of étalé trajectories in Chap. 6, we make reference to the deeper sheaf concepts of étalé topology and stalk topologies, which we present in Sects. 4.3 and 4.4.

In Chap. 5 we provide our modelling framework for geometric control systems, defining what we shall call "tautological control systems".[3] We first provide the definitions and then give the notion of a trajectory for these systems. We also show that our framework includes the standard framework of Sect. 3.3 as a special case. We carefully establish correspondences between our generalised models, the standard models, and differential inclusion models. Included in this correspondence is a description of the relationships between trajectories for these models, and we prove that for control-affine systems, and for systems with general dependence on control and compact control sets, the trajectory equivalence is exact.

In Chap. 6 we present the notion of an étalé system. This extends the basic framework of Chap. 5 to one where the sheaf structure really assumes a prominent rôle. While the punchline of this chapter is the quite natural notion of an étalé trajectory, it is probably the case that the aspects of the tautological control system framework that are touched upon in this chapter will be those that have the most to do with the future utility of the framework.

What is presented in this work is the result of initial explorations of a modelling framework for geometric control theory. We certainly have not fully fleshed out all parts of this framework ourselves. In the closing chapter, Chap. 7, we outline places where there is ongoing and obvious further work to be done. In this chapter we intend to illustrate that the tautological control system framework *naturally* reveals elements of system structure that have hitherto been obscured, and permits the use of tools that have hitherto not been used in control theory.

1.4 Notation, Conventions, and Background

In this section we overview what is needed to read the monograph. We do use a lot of specialised material in essential ways, and we certainly do not review this comprehensively. Instead, we simply provide a few facts, the notation we shall use, and recommended sources. Throughout the work we have tried to include precise references to material needed so that a reader possessing enthusiasm and lacking background can begin to chase down all of the ideas upon which we rely.

We shall use the slightly unconventional, but perfectly rational, notation of writing $A \subseteq B$ to denote set inclusion, and when we write $A \subset B$ we mean that $A \subseteq B$ and $A \neq B$. By id_A we denote the identity map on a set A. For a product $\prod_{i\in I} X_i$ of sets, $\mathrm{pr}_j \colon \prod_{i\in I} X_i \to X_j$ is the projection onto the jth component. We shall have occasion to talk about set-valued maps. If X and Y are sets and Φ is a set-valued map from X to Y, i.e., $\Phi(x)$ is a subset of Y, we shall write $\Phi\colon X \twoheadrightarrow Y$. By $\mathbb{Z}$ we

[3] The terminology "tautological" arises from two different attributes of our framework. First of all, when one makes the natural connection from our systems to standard control systems, we encounter the identity map (Example 5.2–2). Second, in our framework we prove that the only pure feedback transformation is the identity transformation (cf. Proposition 5.39).

denote the set of integers, with $\mathbb{Z}_{\geq 0}$ denoting the set of nonnegative integers and $\mathbb{Z}_{>0}$ denoting the set of positive integers. We denote by $\mathbb{R}$ the set of real numbers. By $\mathbb{R}_{\geq 0}$ we denote the set of nonnegative real numbers and by $\mathbb{R}_{>0}$ the set of positive real numbers. The set of complex numbers is denoted by $\mathbb{C}$.

For a topological space $\mathcal{X}$ and $A \subseteq \mathcal{X}$, $\mathrm{int}(A)$ denotes the interior of A and $\mathrm{cl}(A)$ denotes the closure of A. Neighbourhoods will always be open sets. The support of a continuous function f (or any other kind of object for which it makes sense to have a value "zero") is denoted by $\mathrm{supp}(f)$.

Elements of $\mathbb{R}^n$ are typically denoted with a bold font, e.g., "$\mathbf{x}$". Similarly, matrices are written using a bold font, e.g., "$\mathbf{A}$". By $\|\cdot\|$ we denote the Euclidean norm for $\mathbb{R}^n$ or $\mathbb{C}^n$. By $\mathsf{B}(r, \mathbf{x}) \subseteq \mathbb{R}^n$ we denote the open ball of radius r and centre $\mathbf{x}$. In like manner, $\overline{\mathsf{B}}(r, \mathbf{x})$ denotes the closed ball.

If $\mathcal{U} \subseteq \mathbb{R}^n$ is open and if $\Phi: \mathcal{U} \to \mathbb{R}^m$ is differentiable at $\mathbf{x} \in \mathcal{U}$, we denote its derivative by $\boldsymbol{D}\boldsymbol{\Phi}(\mathbf{x})$. Higher-order derivatives, when they exist, are denoted by $\boldsymbol{D}^r \Phi(\mathbf{x})$, r being the order of differentiation.

If V is a $\mathbb{R}$-vector space and if $A \subseteq \mathsf{V}$, we denote by $\mathrm{conv}(A)$ the convex hull of A, by which we mean the set of all convex combinations of elements of A.

By λ we denote Lebesgue measure. If $I \subseteq \mathbb{R}$ is an interval and if $A \subseteq \mathbb{R}$, by $\mathrm{L}^1(I; A)$ we denote the set of Lebesgue integrable A-valued functions on I. By $\mathrm{L}^1_{\mathrm{loc}}(I; A)$ we denote the A-valued locally integrable functions on I, i.e., those functions whose restrictions to compact subintervals are integrable. In like manner, we denote by $\mathrm{L}^\infty(I; A)$ and $\mathrm{L}^\infty_{\mathrm{loc}}(I; A)$ the essentially bounded A-valued functions and the locally essentially bounded A-valued functions, respectively.

For an interval I and a topological space $\mathcal{X}$, a curve $\gamma: I \to \mathcal{X}$ is ***measurable*** if $\gamma^{-1}(\mathcal{O})$ is Lebesgue measurable for every open $\mathcal{O} \subseteq \mathcal{X}$. By $\mathrm{L}^{\mathrm{cpt}}(I; \mathcal{X})$ we denote the measurable curves $\gamma: I \to \mathcal{X}$ for which there exists a compact set $K \subseteq \mathcal{X}$ with

$$\lambda(\{t \in I \mid \gamma(t) \notin K\}) = 0,$$

i.e., $\mathrm{L}^{\mathrm{cpt}}(I; \mathcal{X})$ is the set of ***essentially bounded*** curves. By $\mathrm{L}^{\mathrm{cpt}}_{\mathrm{loc}}(I; \mathcal{X})$ we denote the ***locally essentially bounded*** curves, meaning those measurable curves whose restrictions to compact subintervals are essentially bounded.

Our differential geometric conventions mostly follow [1]. Whenever we write "manifold", we mean "second-countable Hausdorff manifold". This implies, in particular, that manifolds are assumed to be metrisable [1, Corollary 5.5.13]. If we use the letter "n" without mentioning what it is, it is the dimension of the connected component of the manifold M with which we are working at that time. The tangent bundle of a manifold M is denoted by $\pi_{\mathsf{TM}}: \mathsf{TM} \to \mathsf{M}$ and the cotangent bundle by $\pi_{\mathsf{T}^*\mathsf{M}}: \mathsf{T}^*\mathsf{M} \to \mathsf{M}$. The derivative of a differentiable map $\Phi: \mathsf{M} \to \mathsf{N}$ is denoted by $T\Phi: \mathsf{TM} \to \mathsf{TN}$, with $T_x\Phi = T\Phi|\mathsf{T}_x\mathsf{M}$. If $I \subseteq \mathbb{R}$ is an interval and if $\xi: I \to \mathsf{M}$ is a curve that is differentiable at $t \in I$, we denote the tangent vector field to the curve at t by $\xi'(t) = T_t\xi(1)$.

We will work in both the smooth and real analytic categories. We will also work with finitely differentiable objects, i.e., objects of class C^r for $r \in \mathbb{Z}_{\geq 0}$. (We will also work with Lipschitz objects, but will develop the notation for these in the text.) A good reference for basic real analytic analysis is [44], but a reader wishing to thoroughly understand the real analytic topology we use in this work will need ideas going beyond those from this text, or any other text. We refer to [33] and the references cited there for details. An analytic manifold or mapping will be said to be of ***class*** $\mathbf{C}^{\omega}$. Let $r \in \mathbb{Z}_{\geq 0} \cup \{\infty, \omega\}$. The set of mappings of class C^r between manifolds M and N is denoted by $\mathrm{C}^r(\mathsf{M}; \mathsf{N})$. In particular, $\mathrm{C}^r(\mathsf{M})$ denotes the space of functions of class C^r. The set of sections of a vector bundle $\pi: \mathsf{E} \to \mathsf{M}$ of class C^r is denoted by $\Gamma^r(\mathsf{E})$. Thus, in particular, $\Gamma^r(\mathsf{TM})$ denotes the set of vector fields of class C^r on a manifold M. We shall think of $\Gamma^r(\mathsf{E})$ as a $\mathbb{R}$-vector space with the natural pointwise addition and scalar multiplication operations.

We also work with holomorphic, i.e., complex analytic, manifolds and associated geometric constructions; real analytic geometry, at some level, seems to unavoidably rely on holomorphic geometry. A nice overview of holomorphic geometry, and some of its connections to real analytic geometry, is given in the book [14]. There are many specialised texts on the subject of holomorphic geometry, including the three volumes of Gunning [22–24]. For our purposes, we shall just say the following things. By TM we denote the holomorphic tangent bundle of M. This is the object which, in complex differential geometry, is commonly denoted by $\mathsf{T}^{1,0}\mathsf{M}$. For holomorphic manifolds M and N, we denote by $\mathrm{C}^{\mathrm{hol}}(\mathsf{M}; \mathsf{N})$ the set of holomorphic mappings from M to N, by $\mathrm{C}^{\mathrm{hol}}(\mathsf{M})$ the set of holomorphic functions on M (note that these functions are $\mathbb{C}$-valued, not $\mathbb{R}$-valued, of course), and by $\Gamma^{\mathrm{hol}}(\mathsf{E})$ the space of holomorphic sections of an holomorphic vector bundle $\pi: \mathsf{E} \to \mathsf{M}$. We shall use both the natural $\mathbb{C}$- and, by restriction, $\mathbb{R}$-vector space structures for $\Gamma^{\mathrm{hol}}(\mathsf{E})$.

We shall make reference to mostly elementary ideas from sheaf theory. It will not be necessary to understand this theory deeply, at least not in the present monograph. In particular, a comprehensive understanding of sheaf cohomology is not required, although we do make use of Cartan's Theorems A and B in places. A nice introduction to the use of sheaves in smooth differential geometry can be found in the book of Ramanan [54]. More advanced and comprehensive treatments include [8, 39], and the classic [20]. The discussion of sheaf theory in [60] is also useful. Understandable proofs of Cartan's Theorems in the holomorphic setting can be found in [67]. The real analytic versions of these theorems seem to only be available in the original paper of Cartan [13]. For readers who are expert in sheaf theory, we comment that our reasons for using sheaves are not always the usual ones, so an adjustment of point of view may be required.

We shall make use of locally convex topological vector spaces, and refer to [15, 21, 29, 35, 55, 57] for details. We provide in Sect. 2.1 a rapid overview of the subject.

We shall talk about two flavours of "boundedness", and we will need to carefully discriminate between these. The proper abstract framework for talking about boundedness is supplied by the notion of a “bornology”.[4] Bornologies are less popular than topologies, but a treatment in some generality can be found in [28]. There are two bornologies we consider in this work. One is the ***compact bornology*** for a topological space $\mathcal{X}$ whose bounded sets are the relatively compact sets. The other is the ***von Neumann bornology*** for a locally convex topological vector space V whose bounded sets are those subsets $\mathcal{B} \subseteq \mathsf{V}$ for which, for any neighbourhood $\mathcal{N}$ of $0 \in \mathsf{V}$, there exists $\lambda \in \mathbb{R}_{>0}$ such that $\mathcal{B} \subseteq \lambda\mathcal{N}$. On any locally convex topological vector space we thus have these two bornologies, and generally they are not the same. Indeed, if V is an infinite-dimensional normed vector space, then the compact bornology is strictly contained in the von Neumann bornology. For certain locally convex spaces, however, the compact and von Neumann bornologies agree. An important example of such a class is the nuclear locally convex spaces [52, Proposition 4.47]. We shall merely make use of the fact that some of our spaces are nuclear, and will point this out at the appropriate moments. But, in general, we will have occasion to use both the compact and von Neumann bornologies, and shall make it clear which we mean. This is reflected, for example, in our use of the symbol “L^{∞}” for “essentially bounded” and of the symbol “$\mathrm{L}^{\mathrm{cpt}}$” for “essentially compact-valued”. The latter symbol is not typically used, but in this work we must discriminate between the two notions of boundedness.

References

1. Abraham R, Marsden JE, Ratiu TS (1988) Manifolds, tensor analysis, and applications, 2nd edn. No. 75 in Applied Mathematical Sciences. Springer, Berlin
2. Agrachev AA, Gamkrelidze RV (1978) The exponential representation of flows and the chronological calculus. Math USSR-Sb 107(4):467–532
3. Agrachev AA, Sachkov Y (2004) Control theory from the geometric viewpoint, Encyclopedia of Mathematical Sciences, vol 87. Springer, Berlin
4. Aubin JP, Cellina A (1984) Differential inclusions: set-valued maps and viability theory, Grundlehren der Mathematischen Wissenschaften, vol 264. Springer, Berlin
5. Barbero-Liñán M, Muñoz-Lecanda MC (2009) Geometric approach to Pontryagin’s Maximum Principle. Acta Appl Math 108(2):429–485
6. Bianchini RM, Stefani G (1993) Controllability along a trajectory: a variational approach. SIAM J Control Optim 31(4):900–927
7. Bloch AM (2003) Nonholonomic mechanics and control, Interdisciplinary Applied Mathematics, vol 24. Springer, Berlin
8. Bredon GE (1997) Sheaf theory, 2nd edn. No. 170 in Graduate Texts in Mathematics. Springer, Berlin
9. Brockett RW (1977) Control theory and analytical mechanics. In: Martin C, Hermann R (eds) Geometric control theory. Math Sci Press, Brookline, pp 1–48

[4] A ***bornology*** on a set $\mathcal{S}$ is a family $\mathscr{B}$ of subsets of $\mathcal{S}$, called ***bounded sets***, and satisfying the axioms:

1. $\mathcal{S}$ is covered by bounded sets, i.e., $\mathcal{S} = \cup_{B \in \mathscr{B}} B$;
2. subsets of bounded sets are bounded, i.e., if $B \in \mathscr{B}$ and if $A \subseteq B$, then $A \in \mathscr{B}$;
3. finite unions of bounded sets are bounded, i.e., if $B_1, \dots, B_k \in \mathscr{B}$, then $\cup_{j=1}^{k} B_j \in \mathscr{B}$.

10. Bryant RL, Gardner RB (1993) Control structures. Geometry in nonlinear control and differential inclusions. No. 32 in Banach Center Publications, Polish Academy of Sciences, Institute for Mathematics, Warsaw, pp 111–121
11. Bullo F, Lewis AD (2004) Geometric control of mechanical systems: modeling, analysis, and design for simple mechanical systems. No. 49 in Texts in Applied Mathematics. Springer, Berlin
12. Bus JCP (1984) The Lagrange multiplier rule on manifolds and optimal control of nonlinear systems. SIAM J Control Optim 22(5):740–757
13. Cartan H (1957) Variétés analytiques réelles et variétés analytiques complexes. Bull Soc Math 85:77–99
14. Cieliebak K, Eliashberg Y (2012) From Stein to Weinstein and back: symplectic geometry of affine complex manifolds. No. 59 in American Mathematical Society Colloquium Publications. American Mathematical Society, Providence, RI
15. Conway JB (1985) A course in functional analysis, 2nd edn. No. 96 in Graduate Texts in Mathematics. Springer, Berlin
16. Delgado-Téllez M, Ibort A (2003) A panorama of geometric optimal control theory. Extracta Math 18(2):129–151
17. Filippov AF (1988) Differential equations with discontinuous righthand sides. No. 18 in Mathematics and its Applications (Soviet Series). Kluwer Academic Publishers, Dordrecht
18. Fuller AT (1960) Relay control systems optimized for various performance criteria. In: Proceedings of the First IFAC World Congress. IFAC, Butterworth & Co. Ltd., London, Moscow, pp 510–519
19. Gardner RB (1989) The method of equivalence and its applications. No. 58 in Regional Conference Series in Applied Mathematics. Society for Industrial and Applied Mathematics, Philadelphia, PA
20. Godement R (1958) Topologie algébrique et théorie des faisceaux. No. 13 in Publications de l'Institut de mathématique de l'Université de Strasbourg. Hermann, Paris
21. Groethendieck A (1973) Topological vector spaces. Notes on Mathematics and its Applications. Gordon & Breach Science Publishers, New York
22. Gunning RC (1990) Introduction to holomorphic functions of several variables, vol I: function theory. Wadsworth & Brooks/Cole Mathematics Series. Wadsworth & Brooks/Cole, Belmont
23. Gunning RC (1990) Introduction to holomorphic functions of several variables, vol II: local theory. Wadsworth & Brooks/Cole Mathematics Series. Wadsworth & Brooks/Cole, Belmont
24. Gunning RC (1990) Introduction to holomorphic functions of several variables, vol III: homological theory. Wadsworth & Brooks/Cole Mathematics Series. Wadsworth & Brooks/Cole, Belmont
25. Hermann R (1988) Invariants for feedback equivalence and Cauchy characteristic multifoliations of nonlinear control systems. Acta Appl Math 11(2):123–153
26. Hermann R (1989) Nonlinear feedback control and systems of partial differential equations. Acta Appl Math 17(1):41–94
27. Hermann R, Krener AJ (1977) Nonlinear controllability and observability. Institute of Electrical and Electronics Engineers. Trans Autom Control 22(5):728–740
28. Hogbe-Nlend H (1977) Bornologies and functional analysis. No. 26 in North Holland Mathematical Studies. (Translated from the French by Moscatelli VB). North-Holland, Amsterdam
29. Horváth J (1966) Topological vector spaces and distributions, vol I. Addison Wesley, Reading
30. Isidori A (1995) Nonlinear control systems. Communications and Control Engineering Series, 3rd edn. Springer, Berlin
31. Jafarpour S, Lewis AD (2014) Locally convex topologies and control theory. Submitted to SIAM J Control Optim
32. Jafarpour S, Lewis AD (2014) Real analytic control systems. Submitted to 53rd IEEE Conference on Decision and Control
33. Jafarpour S, Lewis AD (2014) Time-varying vector fields and their flows. To appear in Springer Briefs in Mathematics

34. Jakubczyk B, Respondek W (1980) On linearization of control systems. Bull Acad Polon Sci Sér Sci Math Astronom Phys 28(9–10):517–522
35. Jarchow H (1981) Locally convex spaces. Mathematical Textbooks. Teubner, Leipzig
36. Jurdjevic V (1997) Geometric control theory. No. 51 in Cambridge Studies in Advanced Mathematics. Cambridge University Press, New York
37. Kang W, Krener AJ (1998) Extended quadratic controller normal form and dynamic feedback linearization of nonlinear systems. SIAM J Control Optim 30(6):1319–1337
38. Kang W, Krener AJ (2006) Normal forms of nonlinear control systems. Chaos in automatic control, control engineering. Taylor & Francis, New York
39. Kashiwara M, Schapira P (1990) Sheaves on manifolds. No. 292 in Grundlehren der Mathematischen Wissenschaften. Springer, Berlin
40. Kawski M (1990) High-order small-time local controllability. Nonlinear controllability and optimal control. Monographs and Textbooks in Pure and Applied Mathematics, vol 133. Dekker Marcel Dekker, New York, pp 431–467
41. Kawski M (1999) Controllability via chronological calculus. In: Proceedings of the 38th IEEE conference on decision and control, pp 2920–2926. Institute of Electrical and Electronics Engineers, Phoenix, AZ
42. Kawski M (2006) On the problem whether controllability is finitely determined. In: Proceedings of MTNS '06
43. Khalil HK (1996) Nonlinear systems, 2nd edn. Prentice-Hall, Englewood Cliffs
44. Krantz SG, Parks HR (2002) A primer of real analytic functions. Birkhäuser Advanced Texts, 2nd edn. Birkhäuser, Boston
45. Langerock B (2003) Geometric aspects of the maximum principle and lifts over a bundle map. Acta Appl Math 77(1):71–104
46. Lewis AD (2012) Fundamental problems of geometric control theory. In: Proceedings of the 51st IEEE conference on decision and control, pp 7511–7516. Institute of Electrical and Electronics Engineers, Maui, HI
47. Nagano T (1966) Linear differential systems with singularities and an application to transitive Lie algebras. J Math Soc Jpn 18:398–404
48. Nijmeijer H (1983) Nonlinear multivariable control: a differential geometric approach. Ph.D. thesis, University of Groningen
49. Nijmeijer H, van der Schaft AJ (1982) Controlled invariance for nonlinear systems. Institute of Electrical and Electronics Engineers. Trans Autom Control 27(4):904–914
50. Nijmeijer H, van der Schaft AJ (1990) Nonlinear dynamical control systems. Springer, Berlin
51. Pasillas-Lépine W, Respondek W (2002) Contact systems and corank one involutive subdistributions. Acta Appl Math 69(2):105–128
52. Pietsch A (1969) Nuclear locally convex spaces. No. 66 in Ergebnisse der Mathematik und ihrer Grenzgebiete. Springer, Berlin
53. Polderman JW, Willems JC (1998) Introduction to mathematical systems theory. No. 26 in Texts in Applied Mathematics. Springer, Berlin
54. Ramanan S (2005) Global calculus. No. 65 in Graduate Studies in Mathematics. American Mathematical Society, Providence
55. Rudin W (1991) Functional analysis. International Series in Pure and Applied Mathematics, 2nd edn. McGraw-Hill, New York
56. Sastry S (1999) Nonlinear systems: analysis, stability, and control. No. 10 in Interdisciplinary Applied Mathematics. Springer, Berlin
57. Schaefer HH, Wolff MP (1999) Topological vector spaces, 2nd edn. No. 3 in Graduate Texts in Mathematics. Springer, Berlin
58. Smirnov GV (2002) Introduction to the theory of differential inclusions, Graduate Studies in Mathematics, vol 41. American Mathematical Society, Providence
59. Sontag ED (1998) Mathematical control theory: deterministic finite dimensional systems, 2nd edn. No. 6 in Texts in Applied Mathematics. Springer, Berlin
60. Stacks Project Authors (2014) Stacks Project. http://stacks.math.columbia.edu

61. Sussmann HJ (1978) A sufficient condition for local controllability. SIAM J Control Optim 16(5):790–802
62. Sussmann HJ (1983) Lie brackets and local controllability: a sufficient condition for scalar-input systems. SIAM J Control Optim 21(5):686–713
63. Sussmann HJ (1987) A general theorem on local controllability. SIAM J Control Optim 25(1):158–194
64. Sussmann HJ (1997) An introduction to the coordinate-free maximum principle. In: Jakubczyk B, Respondek W (eds) Geometry of feedback and optimal control. Dekker Marcel Dekker, New York, pp 463–557
65. Sussmann HJ (2002) Needle variations and almost lower semicontinuous differential inclusions. Set-Valued Anal 10(2–3):33–285
66. Sussmann HJ, Jurdjevic V (1972) Controllability of nonlinear systems. J Differ Equ 12:95–116
67. Taylor JL (2002) Several complex variables with connections to algebraic geometry and Lie groups. No. 46 in Graduate Studies in Mathematics. American Mathematical Society, Providence
68. van der Schaft AJ (1983) System theoretic descriptions of physical systems. Ph.D. thesis, University of Groningen
69. Willems JC (1979) System theoretic models for the analysis of physical systems. Ricerche di Automatica 10(2):71–106
70. Willems JC (1986) From time series to linear systems. I. Finite-dimensional linear time invariant systems. Automatica. J IFAC 22(5):561–580
71. Willems JC (1986) From time series to linear systems. II. Exact modelling. Automatica. J IFAC 22(6):675–694
72. Willems JC (1987) From time series to linear systems. III. Approximate modelling. Automatica. J IFAC Int Fed Autom Control 23(1):87–115
73. Willems JC (1991) Paradigms and puzzles in the theory of dynamical systems. Institute of Electrical and Electronics Engineers. IEEE Trans Automat Control 36(3):259–294
74. Willems JC (2007) The behavioral approach to open and interconnected systems. IEEE Control Systems Magazine pp 46–99
75. Wonham WM (1985) Linear multivariable control: a geometric approach, 3rd edn. No. 10 in Applications of Mathematics. Springer, Berlin

Chapter 2
Topologies for Spaces of Vector Fields

In this chapter we review the definitions of the topologies we use for spaces of Lipschitz, finitely differentiable, smooth, and real analytic vector fields. We comment that all topologies we define are locally convex topologies, of which the normed topologies are a special case. However, few of the topologies we define, and none of the interesting ones, are normable. We, therefore, begin with a very rapid review of locally convex topologies, and why they are inevitable in work such as we undertake here.

2.1 An Overview of Locally Convex Topologies for Vector Spaces

In this section we provide a "chatty" overview of locally convex topologies, since this work relies on these in an essential way. The presentation here should be regarded as that of a bare bones introduction, and a reader wishing to understand the subject deeply will wish to refer to references such as [1, 3, 5, 7, 10, 12]. We particularly suggest [10] as a good place to start learning the theory.

2.1.1 Motivation

As mentioned above, few of the topologies we introduce below arise from a norm, and the most interesting ones, e.g., the topologies for spaces of smooth and real analytic vector fields, are decidedly not norm topologies. Let us reflect on why locally convex topologies, such as we use in this work, are natural. Consider first the task of putting a norm on the space $C^0(\mathbb{R})$ of continuous $\mathbb{R}$-valued functions on $\mathbb{R}$. Spaces of continuous functions are in the domain of classical analysis, and so are well-known to the readership of this monograph, e.g., [4, Theorem 7.9]. This is often considered

A.D. Lewis, *Tautological Control Systems*, SpringerBriefs in Control, Automation and Robotics, DOI: 10.1007/978-3-319-08638-5_2

for continuous functions defined on compact spaces, e.g., compact intervals, where the sup-norm suffices to describe the topology in an adequate manner. For continuous functions on noncompact spaces, the sup-norm obviously no longer applies. In such cases, it is common to consider instead functions that "die off" at infinity, as the sup-norm again functions perfectly well for these classes. For the entire space of continuous functions, say $C^0(\mathbb{R})$, the sup-norm is no longer a viable candidate for defining a topology. Instead one can use a family of natural seminorms, one for each compact set $K \subseteq \mathbb{R}$. To be precise, we define

$$p_K(f) = \sup\{|f(x)| \mid x \in K\}.$$

The collection p_K, $K \subseteq \mathbb{R}$ compact, of such seminorms can then be used to define a topology (in a manner that we make precise in Definition 2.2). If one wishes to apply the same reasoning to functions of class C^m, $m \in \mathbb{Z}_{>0}$, we can use the seminorms

$$p_K^m(f) = \sup\{|\boldsymbol{D}^j f(x)| \mid x \in K, j \in \{0, 1, \dots, m\}\}, \qquad K \subseteq \mathbb{R} \text{ compact},$$

on $C^m(\mathbb{R})$, and it is not hard to imagine that this can be used to describe a suitable topology; we define these sorts of topologies precisely below.

By being slightly more clever, one can imagine adapting the above procedure for topologising $C^m(\mathbb{R})$, $m \in \mathbb{Z}_{\geq 0}$, to topologising the space $C^m(\mathsf{M})$ of functions on a smooth manifold M of class C^m. If M is compact, such a space is actually a normed space, since supremums can be taken over the compact set M. If one wishes to topologise the space $C^\infty(\mathsf{M})$ of smooth functions on a smooth manifold, one must account for all derivatives. Let us indicate how to do this for $C^\infty(\mathbb{R})$; we handle the general case in Sect. 2.2.2. For $C^\infty(\mathbb{R})$ we define the seminorms

$$p_{K,m}^\infty(f) = \sup\{|\boldsymbol{D}^j f(x)| \mid x \in K, \ j \in \{0, 1, \dots, m\}\}, \qquad K \subseteq \mathbb{R} \text{ compact}, \ m \in \mathbb{Z}_{\geq 0}.$$

Note that the appropriate adaptation of these seminorms to manifolds will never yield a normed topology, since there will always be infinitely many derivatives to account for.

The point of the preceding motivation is this: topologies defined by families of seminorms arise in natural ways when topologising spaces of functions in differential geometry.

2.1.2 Families of Seminorms and Topologies Defined by These

With the preceding remarks as motivation, let us provide a few precise definitions and state a few facts (without proof) arising from these definitions.

We begin with the notion of a seminorm.

Definition 2.1 (Seminorm) Let $\mathbb{F} \in \{\mathbb{R}, \mathbb{C}\}$ and let V be an $\mathbb{F}$-vector space. A ***seminorm*** for V is a function $p\colon \mathsf{V} \to \mathbb{R}_{\geq 0}$ such that

(i) $p(av) = |a|p(v)$ for $a \in \mathbb{F}$ and $v \in \mathsf{V}$;
(ii) $p(v_1 + v_2) \leq p(v_1) + p(v_2)$. ∘

The reader will note that the missing norm element is the positive definiteness. A moments reflection on the examples above indicates why this omission is necessary. Nonetheless, one can use families of seminorms to define a topology.

Definition 2.2 (The topology defined by a family of seminorms) Let $\mathbb{F} \in \{\mathbb{R}, \mathbb{C}\}$, let V be an $\mathbb{F}$-vector space, and let $\mathscr{P}$ be a family of seminorms for V. The ***topology*** defined by $\mathscr{P}$ is that topology for which the sets

$$\{v \in \mathsf{V} \mid p(v) < r\}, \qquad p \in \mathscr{P},\ r \in \mathbb{R}_{>0},$$

are a subbasis, i.e., open sets in the topology are unions of finite intersections of these sets. The resulting topology is called a ***locally convex*** topology, and an $\mathbb{F}$-vector space with a locally convex topology is called an ***$\mathbb{F}$-locally convex topological vector space***, or simply a ***locally convex space***. ∘

Now let us simply list some attributes of these topologies, referring to the references for details. In the following list, we let $\mathbb{F} \in \{\mathbb{R}, \mathbb{C}\}$ and let U and V be $\mathbb{F}$-locally convex spaces defined by families $\mathscr{Q}$ and $\mathscr{P}$, respectively, of seminorms.

1. The locally convex topology on V is Hausdorff if and only if, for each $v \in \mathsf{V}$, there exists $p \in \mathscr{P}$ such that $p(v) \neq 0$ [10, Theorem 1.37]. Locally convex spaces are often assumed to be Hausdorff, and we shall suppose this to be true for our statements below.
2. Locally convex topologies are translation-invariant, i.e., a neighbourhood basis at 0 translates (by adding v) to a neighbourhood basis at $v \in \mathsf{V}$ [10, Theorem 1.37].
3. We say that a subset $\mathcal{B}$ is ***von Neumann bounded*** if, for any neighbourhood $\mathcal{N}$ of 0, there exists $\lambda \in \mathbb{R}_{>0}$ such that $\mathcal{B} \subseteq \lambda\mathcal{N}$. A subset is von Neumann bounded if and only if $p|\mathcal{B}$ is bounded for every $p \in \mathscr{P}$ [10, Theorem 1.37(b)].
4. A locally convex topology is ***normable*** if it can be defined by a single seminorm which is a norm. A locally convex space is normable if and only if there exists a convex bounded neighbourhood of 0 [10, Theorem 1.39].
5. Compact subsets of locally convex spaces are closed and bounded. However, closed and bounded subsets are not necessarily compact, e.g., closed balls in infinite-dimensional Banach spaces are not compact.
6. Unlike the situation for Banach spaces, there *are* infinite-dimensional locally convex spaces for which closed and bounded sets are compact. An important class of such spaces are the so-called nuclear spaces [9, Proposition 4.47]. A normed space is nuclear if and only if it is finite-dimensional. In this work, many of the spaces we deal with are nuclear.
7. A locally convex space is ***metrisable*** if its topology can be defined by a translation-invariant metric. A locally convex space is metrisable if and only if it can be defined by a countable family of seminorms [10, Remark 1.38(c)].

8. Metrisable topologies are characterised by their convergent sequences. This is a general assertion, following from the fact that metric spaces are first-countable [13, Corollary 10.5]. However, we will encounter locally convex spaces that are not metrisable, and so convergence in such spaces is determined by using nets rather than sequences. Recall that a ***net*** in a set is indexed by points in a directed set, i.e., a partially ordered set $(I, \preceq)$ with the attribute that, given $i_1, i_2 \in I$, there exists $i \in I$ such that $i_1, i_2 \preceq i$. A net $(x_i)_{i\in I}$ in a topological space ***converges*** to x_0 if, for every neighbourhood $\mathcal{O}$ of x_0, there exists $i_0 \in I$ such that $x_i \in \mathcal{O}$ for all $i_0 \preceq i$.
9. A net $(v_i)_{i\in I}$ in V converges to v_0 if and only if, for each $p \in \mathscr{P}$ and each $\varepsilon \in \mathbb{R}_{>0}$, there exists $i_0 \in I$ such that $p(v_i - v_0) < \varepsilon$ for $i_0 \preceq i$.
10. A ***Cauchy net*** in V is a net $(v_i)_{i\in I}$ such that, for each $p \in \mathscr{P}$ and each $\varepsilon \in \mathbb{R}_{>0}$, there exists $i_0 \in I$ such that, if $i_0 \preceq i_1, i_2$, then $p(v_{i_1} - v_{i_2}) < \varepsilon$. A locally convex space is ***complete*** if every Cauchy net converges.
11. A linear map $L : \mathsf{U} \to \mathsf{V}$ is continuous if and only if, for each $p \in \mathscr{P}$, there exist $q_1, \dots, q_k \in \mathscr{Q}$ and $C_1, \dots, C_k \in \mathbb{R}_{>0}$ such that

$$p\,(L(u)) \leq C_1 q_1(u) + \cdots + C_k q_k(u),$$

cf. the discussion in [12, §III.1.1]. We denote by $\mathsf{L}(\mathsf{U}; \mathsf{V})$ the set of continuous linear maps from U to V.

The preceding is all that we shall make direct reference to in this monograph. We mention, however, that our work here relies on the recent work of Jafarpour and Lewis [6], and in this work, especially the development of the topology for spaces of real analytic vector fields, many deep properties of locally convex topologies are used. We shall skirt around these issues, for the most part, in the present monograph.

2.2 Seminorms for Locally Convex Spaces of Vector Fields

We now describe in a little detail the seminorms we use for spaces of vector fields with various regularity, Lipschitz, finitely differentiable, smooth, and real analytic. We also characterise spaces of holomorphic vector fields, because these can often be useful in understanding real analytic vector fields.

While our interest is primarily in spaces of vector fields, it is actually less confusing notationally and conceptually to work instead with spaces of sections of a vector bundle. Thus, throughout this section we will work with a vector bundle $\pi\colon \mathsf{E} \to \mathsf{M}$ that is either smooth, real analytic, or holomorphic, depending on our needs.

2.2.1 Fibre Norms for Jet Bundles

The classes of sections we consider are all characterised by their derivatives in some manner. The appropriate device for considering derivatives of sections is the theory of jet bundles, for which we refer to [11] and [8, §12]. By $\mathsf{J}^m\mathsf{E}$ we denote the vector bundle of m-jets of sections of a smooth vector bundle $\pi\colon \mathsf{E} \to \mathsf{M}$, with $\pi_m\colon \mathsf{J}^m\mathsf{E} \to \mathsf{M}$ denoting the projection. If ξ is a smooth section of E, we denote by $j_m\xi$ the corresponding smooth section of $\mathsf{J}^m\mathsf{E}$.

Sections of $\mathsf{J}^m\mathsf{E}$ should be thought of as sections of E along with their first m derivatives. In a local trivialisation of E, one has the local representatives of the derivatives, order-by-order. Such an order-by-order decomposition of derivatives is not possible globally, however. Nonetheless, following [6, §2.1], we shall mimic this order-by-order decomposition globally using a linear connection ∇^0 on E and an affine connection ∇ on M. First note that ∇ defines a connection on $\mathsf{T}^*\mathsf{M}$ by duality. Also, ∇ and ∇^0 together define a connection ∇^m on $\mathrm{T}^m(\mathsf{T}^*\mathsf{M}) \otimes \mathsf{E}$ by asking that the Leibniz Rule be satisfied for the tensor product. Then, for a smooth section ξ of E, we denote

$$\nabla^{(m)}\xi = \nabla^m \cdots \nabla^1\nabla^0\xi,$$

which is a smooth section of $\mathrm{T}^{m+1}(\mathsf{T}^*\mathsf{M} \otimes \mathsf{E})$. By convention we take $\nabla^{(-1)}\xi = \xi$.

We then have a map

$$\begin{aligned} S^m_{\nabla,\nabla^0}\colon \mathsf{J}^m\mathsf{E} &\to \oplus_{j=0}^m(\mathsf{S}^j\mathsf{T}^*\mathsf{M} \otimes \mathsf{E}) \\ j_m\xi(x) &\mapsto (\xi(x), \mathrm{Sym}_1 \otimes \mathrm{id}_{\mathsf{E}}(\nabla^0\xi)(x), \dots, \mathrm{Sym}_m \otimes \mathrm{id}_{\mathsf{E}}(\nabla^{(m-1)}\xi)(x)), \end{aligned} \tag{2.1}$$

which can be verified to be an isomorphism of vector bundles [6, Lemma 2.1]. Here $\mathrm{Sym}_m\colon \mathrm{T}^m(\mathsf{V}) \to \mathrm{S}^m(\mathsf{V})$ is defined by

$$\mathrm{Sym}_m(v_1 \otimes \cdots \otimes v_m) = \frac{1}{m!}\sum_{\sigma\in\mathfrak{S}_m} v_{\sigma(1)} \otimes \cdots \otimes v_{\sigma(m)}.$$

Now we note that inner products on the components of a tensor product induce in a natural way inner products on the tensor product [6, Lemma 2.2]. Thus, if we suppose that we have a fibre metric $\mathbb{G}_0$ on E and a Riemannian metric $\mathbb{G}$ on M, there is induced a natural fibre metric $\mathbb{G}_m$ on $\mathrm{T}^m(\mathsf{T}^*\mathsf{M}) \otimes \mathsf{E}$ for each $m \in \mathbb{Z}_{\geq 0}$. We then define a fibre metric $\overline{\mathbb{G}}_m$ on $\mathsf{J}^m\mathsf{E}$ by

$$\begin{aligned} &\overline{\mathbb{G}}_m(j_m\xi(x), j_m\eta(x)) \\ &\quad = \sum_{j=0}^m \mathbb{G}_j\Big(\frac{1}{j!}\mathrm{Sym}_j \otimes \mathrm{id}_{\mathsf{E}}(\nabla^{(j-1)}\xi)(x), \frac{1}{j!}\mathrm{Sym}_j \otimes \mathrm{id}_{\mathsf{E}}(\nabla^{(j-1)}\eta)(x)\Big). \end{aligned}$$

(The factorials are required to make things work out with the real analytic topology.) The corresponding fibre norm we denote by $\|\cdot\|_{\overline{\mathbb{G}}_m}$.

2.2.2 Seminorms for Spaces of Smooth Vector Fields

Let $\pi\colon \mathsf{E} \to \mathsf{M}$ be a smooth vector bundle. Using the fibre norms from the preceding section, it is a straightforward matter to define appropriate seminorms that prescribe the locally convex topology for $\Gamma^\infty(\mathsf{E})$. For $K \subseteq \mathsf{M}$ compact and for $m \in \mathbb{Z}^m_{\geq 0}$, define a seminorm $p^\infty_{K,m}$ on $\Gamma^\infty(\mathsf{E})$ by

$$p^\infty_{K,m}(\xi) = \sup\{\|j_m\xi(x)\|_{\overline{\mathbb{G}}_m} \mid x \in K\}.$$

The family of seminorms $p^\infty_{K,m}$, $K \subseteq \mathsf{M}$ compact, $m \in \mathbb{Z}_{\geq 0}$, defines a locally convex topology, called the **C$^\infty$-*topology***,[1] with the following properties:

1. it is Hausdorff, metrisable, and complete, i.e., it is a Fréchet topology;
2. it is separable;
3. it is nuclear;
4. it is characterised by the sequences converging to zero, which are the sequences $(\xi_j)_{j\in\mathbb{Z}_{>0}}$ such that, for each $K \subseteq \mathsf{M}$ and $m \in \mathbb{Z}_{\geq 0}$, the sequence $(j_m\xi_j|K)_{j\in\mathbb{Z}_{>0}}$ converges uniformly to zero.

In this paper we shall not make reference to other properties of the C$^\infty$-topology, but we mention that there are other properties that play an important rôle in the results in Chap. 3. For these details, and for references where the above properties are proved, we refer to [6, §3.2].

2.2.3 Seminorms for Spaces of Finitely Differentiable Vector Fields

We again take $\pi\colon \mathsf{E} \to \mathsf{M}$ to be a smooth vector bundle, and we fix $m \in \mathbb{Z}_{\geq 0}$. For the space $\Gamma^m(\mathsf{E})$ of m-times continuously differentiable sections, we define seminorms p^m_K, $K \subseteq \mathsf{M}$ compact, for $\Gamma^m(\mathsf{E})$ by

$$p^m_K(\xi) = \sup\{\|j_m\xi(x)\|_{\overline{\mathbb{G}}_m} \mid x \in K\}.$$

The locally convex topology defined by the family of seminorms p^m_K, $K \subseteq \mathsf{M}$ compact, we call the **C^m-*topology***, and it has the following properties:

1. it is Hausdorff, metrisable, and complete, i.e., it is a Fréchet topology;
2. it is separable;

[1] This is actually not a very good name. A better name, and the name used by Jafarpour and Lewis [6], would be the "smooth compact-open topology". However, we wish to keep things simple here, and also use notation that is common between regularity classes.

3. it is characterised by the sequences converging to zero, which are the sequences $(\xi_j)_{j\in\mathbb{Z}_{>0}}$ such that, for each $K \subseteq \mathsf{M}$, the sequence $(j_m\xi_j|K)_{j\in\mathbb{Z}_{>0}}$ converges uniformly to zero;
4. if M is compact, then p^m_{M} is a norm that gives the C^m-topology.

As with the C^∞-topology, we refer to [6, §3.4] for details.

2.2.4 Seminorms for Spaces of Lipschitz Vector Fields

In this section we again work with a smooth vector bundle $\pi : \mathsf{E} \to \mathsf{M}$. In defining the fibre metrics from Sect. 2.2.1, for the Lipschitz topologies the affine connection ∇ is required to be the Levi-Civita connection for the Riemannian metric $\mathbb{G}$ and the linear connection ∇^0 is required to be $\mathbb{G}_0$-orthogonal. While Lipschitz vector fields are often used, spaces of Lipschitz vector fields are not. Nonetheless, one may define seminorms for spaces of Lipschitz vector fields rather analogous to those defined above in the smooth and finitely differentiable cases. Let $m \in \mathbb{Z}_{\geq 0}$. By $\Gamma^{m+\mathrm{lip}}(\mathsf{E})$ we denote the space of sections of E that are m-times continuously differentiable and whose m-jet is locally Lipschitz. (One can think of this in coordinates, but Jafarpour and Lewis [6] provide geometric definitions, if the reader is interested.) If a section ξ is of class $\mathrm{C}^{m+\mathrm{lip}}$, then, by Rademacher's Theorem [2, Theorem 3.1.6], its $(m+1)$st derivative exists almost everywhere. Thus we define

$$\begin{aligned}\operatorname{dil} j_m\xi(x) = \inf\{\sup\{\|\nabla^{[m]}_{v_y} j_m\xi\|_{\overline{\mathbb{G}}_m} \mid y \in \mathrm{cl}(\mathcal{U}),\ \|v_y\|_{\mathbb{G}} = 1,\\ j_m\xi \text{ differentiable at } y\} \mid \mathcal{U} \text{ is a relatively compact neighbourhood of } x\},\end{aligned}$$

which is the ***local sectional dilatation*** of ξ. Here $\nabla^{[m]}$ is the connection in $\mathsf{J}^m\mathsf{E}$ defined by the decomposition (2.1). Let $K \subseteq \mathsf{M}$ be compact and define

$$\lambda^m_K(\xi) = \sup\{\operatorname{dil} j_m\xi(x) \mid x \in K\}$$

for $\xi \in \Gamma^{m+\mathrm{lip}}(\mathsf{E})$. We can then define a seminorm $p^{m+\mathrm{lip}}_K$ on $\Gamma^{m+\mathrm{lip}}(\mathsf{E})$ by

$$p^{m+\mathrm{lip}}_K(\xi) = \max\{\lambda^m_K(\xi), p^m_K(\xi)\}.$$

The family of seminorms $p^{m+\mathrm{lip}}_K$, $K \subseteq \mathsf{M}$ compact, defines a locally convex topology for $\Gamma^{m+\mathrm{lip}}(\mathsf{E})$, which we call the $\mathbf{C}^{m+\mathrm{lip}}$***-topology***, having the following attributes:

1. it is Hausdorff, metrisable, and complete, i.e., it is a Fréchet topology;
2. it is separable;
3. it is characterised by the sequences converging to zero, which are the sequences $(\xi_j)_{j\in\mathbb{Z}_{>0}}$ such that, for each $K \subseteq \mathsf{M}$, the sequence $(j_m\xi_j|K)_{j\in\mathbb{Z}_{>0}}$ converges uniformly to zero in both seminorms λ^m_K and p^m_K;
4. if M is compact, then $p^{m+\mathrm{lip}}_{\mathsf{M}}$ is a norm that gives the $\mathrm{C}^{m+\mathrm{lip}}$-topology.

We refer to [6, §3.5] for details.

2.2.5 Seminorms for Spaces of Holomorphic Vector Fields

Now we consider an holomorphic vector bundle $\pi\colon \mathsf{E} \to \mathsf{M}$ and denote by $\Gamma^{\mathrm{hol}}(\mathsf{E})$ the space of holomorphic sections of E. We let $\mathbb{G}$ be an Hermitian metric on the vector bundle and denote by $\|\cdot\|_{\mathbb{G}}$ the associated fibre norm. For $K \subseteq \mathsf{M}$ compact, denote by p_K^{hol} the seminorm

$$p_K^{\mathrm{hol}}(\xi) = \sup\{\|\xi(z)\|_{\mathbb{G}} \mid z \in K\}$$

on $\Gamma^{\mathrm{hol}}(\mathsf{E})$. The family of seminorms p_K^{hol}, $K \subseteq \mathsf{M}$ compact, defines a locally convex topology for $\Gamma^{\mathrm{hol}}(\mathsf{E})$ that we call the **C^{hol}*-topology***. This topology has the following properties:

1. it is Hausdorff, metrisable, and complete, i.e., it is a Fréchet topology;
2. it is separable;
3. it is nuclear;
4. it is characterised by the sequences converging to zero, which are the sequences $(\xi_j)_{j\in\mathbb{Z}_{>0}}$ such that, for each $K \subseteq \mathsf{M}$, the sequence $(\xi_j|K)_{j\in\mathbb{Z}_{>0}}$ converges uniformly to zero;
5. if M is compact, then $p_{\mathsf{M}}^{\mathrm{hol}}$ is a norm that gives the $\mathrm{C}^{\mathrm{hol}}$-topology.

We refer to [6, §4.2] and the references therein for details about the $\mathrm{C}^{\mathrm{hol}}$-topology.

2.2.6 Seminorms for Spaces of Real Analytic Vector Fields

The topologies described above for spaces of smooth, finitely differentiable, Lipschitz, and holomorphic sections of a vector bundle are quite simple to understand in terms of their converging sequences. The topology one considers for real analytic sections does not have this attribute. There is a bit of a history to the characterisation of real analytic topologies, and we refer to [6, §5] for *four* equivalent characterisations of the real analytic topology for the space of real analytic sections of a vector bundle. Here we will give the most elementary of these definitions to state, although it is probably not the most practical definition. In practice, it is probably best to somehow complexify and use the holomorphic topology; we give instances of this in Theorems 3.9 and 3.17 below.

In this section we let $\pi\colon \mathsf{E} \to \mathsf{M}$ be a real analytic vector bundle and let $\Gamma^{\omega}(\mathsf{E})$ be the space of real analytic sections. One can show that there exist a real analytic linear connection ∇^0 on E, a real analytic affine connection ∇ on M, a real analytic fibre metric on E, and a real analytic Riemannian metric on M [6, Lemma 2.3]. Thus we can define real analytic fibre metrics $\overline{\mathbb{G}}_m$ on the jet bundles $\mathsf{J}^m\mathsf{E}$ as in Sect. 2.2.1.

To define seminorms for $\Gamma^{\omega}(\mathsf{E})$, let $c_0(\mathbb{Z}_{\geq 0}; \mathbb{R}_{>0})$ denote the space of sequences in $\mathbb{R}_{>0}$, indexed by $\mathbb{Z}_{\geq 0}$, and converging to zero. We shall denote a typical element of $c_0(\mathbb{Z}_{\geq 0}; \mathbb{R}_{>0})$ by $\mathbf{a} = (a_j)_{j \in \mathbb{Z}_{\geq 0}}$. Now, for $K \subseteq \mathsf{M}$ and $\mathbf{a} \in c_0(\mathbb{Z}_{\geq 0}; \mathbb{R}_{>0})$, we define a seminorm $p^{\omega}_{K,\mathbf{a}}$ for $\Gamma^{\omega}(\mathsf{E})$ by

$$p^{\omega}_{K,\mathbf{a}}(\xi) = \sup\{a_0 a_1 \cdots a_m \| j_m \xi(x) \|_{\overline{\mathbb{G}}_m} \mid x \in K, \ m \in \mathbb{Z}_{\geq 0}\}.$$

The family of seminorms $p^{\omega}_{K,\mathbf{a}}$, $K \subseteq \mathsf{M}$ compact, $\mathbf{a} \in c_0(\mathbb{Z}_{\geq 0}; \mathbb{R}_{>0})$, defines a locally convex topology on $\Gamma^{\omega}(\mathsf{E})$ that we call the $\mathbf{C}^{\omega}$***-topology***. This topology has the following attributes:

1. it is Hausdorff and complete;
2. it is not metrisable (and so it not a Fréchet topology);
3. it is separable;
4. it is nuclear.

We shall generally avoid dealing with the rather complicated structure of this topology, and shall be able to do what we need by just working with the seminorms. That this is possible is one of the main contributions of the work [6].

2.2.7 Summary and Notation

In the real case, the degrees of regularity are ordered according to

$$\mathrm{C}^0 \supset \mathrm{C}^{\mathrm{lip}} \supset \mathrm{C}^1 \supset \cdots \supset \mathrm{C}^m \supseteq \mathrm{C}^{m+\mathrm{lip}} \supset \mathrm{C}^{m+1} \supset \cdots \supset \mathrm{C}^{\infty} \supset \mathrm{C}^{\omega},$$

and in the complex case the ordering is the same, of course, but with an extra $\mathrm{C}^{\mathrm{hol}}$ on the right. Sometimes it will be convenient to write $\nu + \mathrm{lip}$ for $\nu \in \{\mathbb{Z}_{\geq 0}, \infty, \omega\}$, and in doing this we adopt the obvious convention that $\infty + \mathrm{lip} = \infty$ and $\omega + \mathrm{lip} = \omega$.

Where possible, we will state definitions and results for all regularity classes at once. To do this, we will let $m \in \mathbb{Z}_{\geq 0}$ and $m' \in \{0, \mathrm{lip}\}$, and consider the regularity classes $\nu \in \{m + m', \infty, \omega\}$. In such cases we shall require that the underlying manifold be of class "C^r, $r \in \{\infty, \omega\}$, as required". This has the obvious meaning, namely that we consider class C^{ω} if $\nu = \omega$ and class C^{∞} otherwise. Proofs will typically break into the four cases $\nu = \infty$, $\nu = m$, $\nu = m + \mathrm{lip}$, and $\nu = \omega$. In some cases there is a structural similarity in the way arguments are carried out, so we will sometimes do all cases at once. In doing this, we will, for $K \subseteq \mathsf{M}$ be compact, for $k \in \mathbb{Z}_{\geq 0}$, and for $\mathbf{a} \in c_0(\mathbb{Z}_{\geq 0}; \mathbb{R}_{>0})$, denote

$$p_K = \begin{cases} p^{\infty}_{K,k}, & \nu = \infty, \\ p^{m}_{K}, & \nu = m, \\ p^{m+\mathrm{lip}}_{K}, & \nu = m + \mathrm{lip}, \\ p^{\omega}_{K,\mathbf{a}}, & \nu = \omega. \end{cases} \tag{2.2}$$

The convenience and brevity more than make up for the slight loss of preciseness in this approach.

References

1. Conway JB (1985) A course in functional analysis, 2nd edn. No. 96 in Graduate Texts in Mathematics. Springer, New York
2. Federer H (1996) Geometric measure theory. Classics in Mathematics. Springer, New York (reprint of 1969 edition)
3. Groethendieck A (1973) Topological vector spaces. Notes on Mathematics and its Applications. Gordon & Breach Science Publishers, New York
4. Hewitt E, Stromberg K (1975) Real and abstract analysis. No. 25 in Graduate Texts in Mathematics. Springer, New York
5. Horváth J (1966) Topological vector spaces and distributions, vol I. Addison Wesley, Reading, MA
6. Jafarpour S, Lewis AD (2014) Time-varying vector fields and their flows. To appear in Springer Briefs in Mathematics
7. Jarchow H (1981) Locally convex spaces. Mathematical Textbooks. Teubner, Leipzig
8. Kolář I, Michor PW, Slovák J (1993) Natural operations in differential geometry. Springer, New York
9. Pietsch A (1969) Nuclear locally convex spaces. No. 66 in Ergebnisse der Mathematik und ihrer Grenzgebiete. Springer, New York
10. Rudin W (1991) Functional Analysis, 2nd edn. International Series in Pure and Applied Mathematics. McGraw-Hill, New York
11. Saunders DJ (1989) The geometry of jet bundles. No. 142 in Lecture Note Series in London Mathematical Society. Cambridge University Press, New York
12. Schaefer HH, Wolff MP (1999) Topological vector spaces, 2 edn. No. 3 in Graduate Texts in Mathematics. Springer, New York
13. Willard S (2004) General topology. Dover Publications, Inc., New York (Reprint of 1970 Addison-Wesley edition)

Chapter 3
Time-Varying Vector Fields and Control Systems

We now turn to utilising the locally convex topologies from the preceding chapter to characterise time-varying vector fields and control systems. We shall see that the use of locally convex topologies allows for a comprehensive and unified treatment of these notions, and allows one to understand in a deep way their structure in a way which has hitherto not been possible. This is especially true in the real analytic case, where we describe new results of Jafarpour and Lewis [6, 7] on the structure of time-varying vector fields and control systems depending on state in a real analytic manner.

The ideas we present here seem to first appear in the work of Agrachev and Gamkrelidze [1]; see also the presentation of Agrachev and Sachkov [2]. Our treatment is different in a few ways. First of all, we make use of connections and jet bundles, aided by (2.1). In [1] the presentation is developed on Euclidean spaces, and so the geometric treatment we give here is not necessary. (One way of understanding why it is not necessary is that Euclidean space has a canonical flat connection in which the decomposition of (2.1) becomes the usual decomposition of derivatives by their order.) In [2] the treatment is on manifolds, and the seminorms are defined by an embedding of the manifold in Euclidean space by Whitney's Embedding Theorem [9]. In [2] consideration is only given to smooth vector fields. In [1] the real analytic case is also considered, but in a rather restricted setting, i.e., to real analytic vector fields on real Euclidean space admitting a bounded holomorphic extension to a strip of fixed width in complex Euclidean space. Our presentation is far more general and geometric. Moreover, we also include the finitely differentiable and Lipschitz classes. Also, Agrachev and Sachkov [2] use what Jafarpour and Lewis [7] call the "weak-$\mathscr{L}$ topology" in the case of vector fields. This is shown in [7, Theorems 3.5, 3.8, 3.14, 4.5, and 5.8] to be the same as the usual topology for all regularity classes. In the work of Sussmann [8] a similarly styled presentation of time-varying vector fields is given. As with the work of Agrachev and coauthors, Sussmann relies on the "weak-$\mathscr{L}$" characterisation of time-dependence. Moreover, the locally convex topology for $\Gamma^\infty(\mathsf{TM})$ is not explicitly considered, although it is implicit in Sussmann's constructions. In any case, all approaches can be tediously shown to be equivalent (in the smooth case) once the relationships are understood.

A.D. Lewis, *Tautological Control Systems*, SpringerBriefs in Control, Automation and Robotics, DOI: 10.1007/978-3-319-08638-5_3

Apart from the fact that we handle all common regularity classes, an advantage of the approach we use here is that it does not require coordinate charts or embeddings to write the seminorms, and it makes the seminorms explicit, rather than implicitly present. The disadvantage of our approach is the added machinery and complication of connections and our jet bundle decomposition.

As in Sect. 2.2, it is convenient to work with sections of a general vector bundle $\pi : \mathsf{E} \to \mathsf{M}$, rather than just with vector fields.

3.1 Time-Varying Vector Fields

The work of Jafarpour and Lewis [7] is concerned with time-varying vector fields with measurable time dependence. In that work, a comprehensive and consistent theory for such vector fields, with varying regularity in state, is developed. Thus, for $m \in \mathbb{Z}_{\geq 0}$, $m' \in \{0, \text{lip}\}$, for $\nu \in \{m + m', \infty, \omega\}$, and for an interval $\mathbb{T} \subseteq \mathbb{R}$, characterisations are given for classes of time-varying vector fields, denoted by $\mathrm{LI}\Gamma^{\nu}(\mathbb{T}; \mathsf{TM})$. There are two equivalent ways to present these classes of vector fields: (1) by directly prescribing the joint pointwise conditions on state and time in each regularity class; (2) using the C^{ν}-topologies. We shall present the definitions in the former setting, and also state without proof the equivalence of the two seemingly unrelated characterisations.

3.1.1 Time-Varying Smooth Vector Fields

Throughout this section we will work with a smooth vector bundle $\pi : \mathsf{E} \to \mathsf{M}$ with a linear connection ∇^0 on E, an affine connection ∇ on M, a fibre metric $\mathbb{G}_0$ on E, and a Riemannian metric $\mathbb{G}$ on M. This defines the fibre norms $\|\cdot\|_{\overline{\mathbb{G}}_m}$ on $\mathsf{J}^m\mathsf{E}$ and seminorms $p^{\infty}_{K,m}$, $K \subseteq \mathsf{M}$ compact, $m \in \mathbb{Z}_{\geq 0}$, on $\Gamma^{\infty}(\mathsf{E})$ as in Sect. 2.2.2.

Definition 3.1 (Smooth Carathéodory section). Let $\pi : \mathsf{E} \to \mathsf{M}$ be a smooth vector bundle and let $\mathbb{T} \subseteq \mathbb{R}$ be an interval. A ***Carathéodory section of class*** C^{∞} of E is a map $\xi : \mathbb{T} \times \mathsf{M} \to \mathsf{E}$ with the following properties:

(i) $\xi(t, x) \in \mathsf{E}_x$ for each $(t, x) \in \mathbb{T} \times \mathsf{M}$;
(ii) for each $t \in \mathbb{T}$, the map $\xi_t : \mathsf{M} \to \mathsf{E}$ defined by $\xi_t(x) = \xi(t, x)$ is of class C^{∞};
(iii) for each $x \in \mathsf{M}$, the map $\xi^x : \mathbb{T} \to \mathsf{E}$ defined by $\xi^x(t) = \xi(t, x)$ is Lebesgue measurable.

We shall call $\mathbb{T}$ the ***time-domain*** for the section. By $\mathrm{CF}\Gamma^{\infty}(\mathbb{T}; \mathsf{E})$ we denote the set of Carathéodory sections of class C^{∞} of E. ∘

Note that the curve $t \mapsto \xi(t, x)$ is in the finite-dimensional vector space E_x, and so Lebesgue measurability of this is unambiguously defined, e.g., by choosing a basis and asking for Lebesgue measurability of the components with respect to this basis.

Now we put some conditions on the time dependence of the derivatives of the section.

Definition 3.2 (Locally integrally C^∞-bounded and locally essentially C^∞-bounded sections). Let $\pi: \mathsf{E} \to \mathsf{M}$ be a smooth vector bundle and let $\mathbb{T} \subseteq \mathbb{R}$ be an interval. A Carathéodory section $\xi: \mathbb{T} \times \mathsf{M} \to \mathsf{E}$ of class C^∞ is

(i) ***locally integrally*** C^∞***-bounded*** if, for every compact set $K \subseteq \mathsf{M}$ and every $m \in \mathbb{Z}_{\geq 0}$, there exists $g \in L^1_{\mathrm{loc}}(\mathbb{T}; \mathbb{R}_{\geq 0})$ such that

$$\|j_m\xi_t(x)\|_{\overline{\mathbb{G}}_m} \leq g(t), \qquad (t, x) \in \mathbb{T} \times K,$$

and is

(ii) ***locally essentially*** C^∞***-bounded*** if, for every compact set $K \subseteq \mathsf{M}$ and every $m \in \mathbb{Z}_{\geq 0}$, there exists $g \in L^\infty_{\mathrm{loc}}(\mathbb{T}; \mathbb{R}_{\geq 0})$ such that

$$\|j_m\xi_t(x)\|_{\overline{\mathbb{G}}_m} \leq g(t), \qquad (t, x) \in \mathbb{T} \times K.$$

The set of locally integrally C^∞-bounded sections of E with time-domain $\mathbb{T}$ is denoted by $\mathrm{LI}\Gamma^\infty(\mathbb{T}, \mathsf{E})$ and the set of locally essentially C^∞-bounded sections of E with time-domain $\mathbb{T}$ is denoted by $\mathrm{LB}\Gamma^\infty(\mathbb{T}; \mathsf{E})$. ∘

Note that $\mathrm{LB}\Gamma^\infty(\mathbb{T}; \mathsf{M}) \subseteq \mathrm{LI}\Gamma^\infty(\mathbb{T}; \mathsf{M})$, precisely because locally essentially bounded functions (in the usual sense) are locally integrable (in the usual sense).

3.1.2 *Time-Varying Finitely Differentiable and Lipschitz Vector Fields*

In this section, so as to be consistent with our definition of Lipschitz seminorms in Sect. 2.2.4, we suppose that the affine connection ∇ on M is the Levi-Civita connection for the Riemannian metric $\mathbb{G}$ and that the vector bundle connection ∇^0 in E is $\mathbb{G}_0$-orthogonal.

Definition 3.3 (Finitely differentiable or Lipschitz Carathéodory section). Let $\pi: \mathsf{E} \to \mathsf{M}$ be a smooth vector bundle and let $\mathbb{T} \subseteq \mathbb{R}$ be an interval. Let $m \in \mathbb{Z}_{\geq 0}$ and let $m' \in \{0, \mathrm{lip}\}$. A **Carathéodory section of class $C^{m+m'}$** of E is a map $\xi: \mathbb{T}\times\mathsf{M} \to \mathsf{E}$ with the following properties:

(i) $\xi(t, x) \in \mathsf{E}_x$ for each $(t, x) \in \mathbb{T} \times \mathsf{M}$;
(ii) for each $t \in \mathbb{T}$, the map $\xi_t: \mathsf{M} \to \mathsf{E}$ defined by $\xi_t(x) = \xi(t, x)$ is of class $C^{m+m'}$;
(iii) for each $x \in \mathsf{M}$, the map $\xi^x: \mathbb{T} \to \mathsf{E}$ defined by $\xi^x(t) = \xi(t, x)$ is Lebesgue measurable.

We shall call $\mathbb{T}$ the ***time-domain*** for the section. By $\mathrm{CF}\Gamma^{m+m'}(\mathbb{T}; \mathsf{E})$ we denote the set of Carathéodory sections of class $C^{m+m'}$ of E. ∘

Now we put some conditions on the time dependence of the derivatives of the section.

Definition 3.4 (Locally integrally $\mathrm{C}^{m+m'}$-bounded and locally essentially $\mathrm{C}^{m+m'}$-bounded sections). Let $\pi: \mathsf{E} \to \mathsf{M}$ be a smooth vector bundle and let $\mathbb{T} \subseteq \mathbb{R}$ be an interval. Let $m \in \mathbb{Z}_{\geq 0}$ and let $m' \in \{0, \mathrm{lip}\}$. A Carathéodory section $\xi: \mathbb{T} \times \mathsf{M} \to \mathsf{E}$ of class $\mathrm{C}^{m+m'}$ is

(i) ***locally integrally*** $\mathbf{C}^{m+m'}$***-bounded*** if:

(a) $m' = 0$: for every compact set $K \subseteq \mathsf{M}$, there exists $g \in \mathrm{L}^1_{\mathrm{loc}}(\mathbb{T}; \mathbb{R}_{\geq 0})$ such that

$$\|j_m \xi_t(x)\|_{\overline{\mathbb{G}}_m} \leq g(t), \qquad (t, x) \in \mathbb{T} \times K;$$

(b) $m' = \mathrm{lip}$: for every compact set $K \subseteq \mathsf{M}$, there exists $g \in \mathrm{L}^1_{\mathrm{loc}}(\mathbb{T}; \mathbb{R}_{\geq 0})$ such that

$$\mathrm{dil}\, j_m \xi_t(x), \|j_m \xi_t(x)\|_{\overline{\mathbb{G}}_m} \leq g(t), \qquad (t, x) \in \mathbb{T} \times K,$$

and is

(ii) ***locally essentially*** $\mathbf{C}^{m+m'}$***-bounded*** if:

(a) $m' = 0$: for every compact set $K \subseteq \mathsf{M}$, there exists $g \in \mathrm{L}^\infty_{\mathrm{loc}}(\mathbb{T}; \mathbb{R}_{\geq 0})$ such that

$$\|j_m \xi_t(x)\|_{\overline{\mathbb{G}}_m} \leq g(t), \qquad (t, x) \in \mathbb{T} \times K;$$

(b) $m' = \mathrm{lip}$: for every compact set $K \subseteq \mathsf{M}$, there exists $g \in \mathrm{L}^\infty_{\mathrm{loc}}(\mathbb{T}; \mathbb{R}_{\geq 0})$ such that

$$\mathrm{dil}\, j_m \xi_t(x), \|j_m \xi_t(x)\|_{\overline{\mathbb{G}}_m} \leq g(t), \qquad (t, x) \in \mathbb{T} \times K.$$

The set of locally integrally $\mathrm{C}^{m+m'}$-bounded sections of E with time-domain $\mathbb{T}$ is denoted by $\mathrm{LI}\Gamma^{m+m'}(\mathbb{T}, \mathsf{E})$ and the set of locally essentially $\mathrm{C}^{m+m'}$-bounded sections of E with time-domain $\mathbb{T}$ is denoted by $\mathrm{LB}\Gamma^{m+m'}(\mathbb{T}; \mathsf{E})$. ∘

3.1.3 Time-Varying Holomorphic Vector Fields

While we are not per se interested in time-varying holomorphic vector fields, our understanding of time-varying real analytic vector fields—in which we are most definitely interested—is connected with an understanding of the holomorphic case, cf. Theorem 3.9.

We begin with definitions that are similar to the smooth and finitely differentiable cases, but which rely on the holomorphic topologies introduced in Sect. 2.2.5. We will consider an holomorphic vector bundle $\pi: \mathsf{E} \to \mathsf{M}$ with an Hermitian fibre metric $\mathbb{G}$. This defines the seminorms p^{hol}_K, $K \subseteq \mathsf{M}$ compact, describing the $\mathrm{C}^{\mathrm{hol}}$-topology for $\Gamma^{\mathrm{hol}}(\mathsf{E})$ as in Sect. 2.2.5.

Let us get started with the definitions.

Definition 3.5 (Holomorphic Carathéodory section). Let $\pi: \mathsf{E} \to \mathsf{M}$ be an holomorphic vector bundle and let $\mathbb{T} \subseteq \mathbb{R}$ be an interval. A ***Carathéodory section of class*** $\mathrm{C}^{\mathrm{hol}}$ of E is a map $\xi: \mathbb{T} \times \mathsf{M} \to \mathsf{E}$ with the following properties:

(i) $\xi(t, z) \in \mathsf{E}_z$ for each $(t, z) \in \mathbb{T} \times \mathsf{M}$;
(ii) for each $t \in \mathbb{T}$, the map $\xi_t: \mathsf{M} \to \mathsf{E}$ defined by $\xi_t(z)$ is of class $\mathrm{C}^{\mathrm{hol}}$;
(iii) for each $z \in \mathsf{M}$, the map $\xi^z: \mathbb{T} \to \mathsf{E}$ defined by $\xi^z(t) = \xi(t, z)$ is Lebesgue measurable.

We shall call $\mathbb{T}$ the ***time-domain*** for the section. By $\mathrm{CF}\Gamma^{\mathrm{hol}}(\mathbb{T}; \mathsf{E})$ we denote the set of Carathéodory sections of class $\mathrm{C}^{\mathrm{hol}}$ of E. ∘

The associated notions for time-dependent sections compatible with the $\mathrm{C}^{\mathrm{hol}}$-topology are as follows.

Definition 3.6 (Locally integrally $\mathrm{C}^{\mathrm{hol}}$-bounded and locally essentially $\mathrm{C}^{\mathrm{hol}}$-bounded sections). Let $\pi: \mathsf{E} \to \mathsf{M}$ be an holomorphic vector bundle and let $\mathbb{T} \subseteq \mathbb{R}$ be an interval. A Carathéodory section $\xi: \mathbb{T} \times \mathsf{M} \to \mathsf{E}$ of class $\mathrm{C}^{\mathrm{hol}}$ is

(i) ***locally integrally*** $\mathrm{C}^{\mathrm{hol}}$***-bounded*** if, for every compact set $K \subseteq \mathsf{M}$, there exists $g \in \mathrm{L}^1_{\mathrm{loc}}(\mathbb{T}; \mathbb{R}_{\geq 0})$ such that

$$\|\xi(t, z)\|_{\mathbb{G}} \leq g(t), \qquad (t, z) \in \mathbb{T} \times K$$

and is

(ii) ***locally essentially*** $\mathrm{C}^{\mathrm{hol}}$***-bounded*** if, for every compact set $K \subseteq \mathsf{M}$, there exists $g \in \mathrm{L}^\infty_{\mathrm{loc}}(\mathbb{T}; \mathbb{R}_{\geq 0})$ such that

$$\|\xi(t, z)\|_{\mathbb{G}} \leq g(t), \qquad (t, z) \in \mathbb{T} \times K.$$

The set of locally integrally $\mathrm{C}^{\mathrm{hol}}$-bounded sections of E with time-domain $\mathbb{T}$ is denoted by $\mathrm{LI}\Gamma^{\mathrm{hol}}(\mathbb{T}, \mathsf{E})$ and the set of locally essentially $\mathrm{C}^{\mathrm{hol}}$-bounded sections of E with time-domain $\mathbb{T}$ is denoted by $\mathrm{LB}\Gamma^{\mathrm{hol}}(\mathbb{T}; \mathsf{E})$. ∘

3.1.4 Time-Varying Real Analytic Vector Fields

Let us now turn to describing real analytic time-varying sections. We thus will consider a real analytic vector bundle $\pi: \mathsf{E} \to \mathsf{M}$ with ∇^0 a real analytic linear connection on E, ∇ a real analytic affine connection on M, $\mathbb{G}_0$ a real analytic fibre metric on E, and $\mathbb{G}$ a real analytic Riemannian metric on M. This defines the seminorms $p^\omega_{K,\mathbf{a}}$, $K \subseteq \mathsf{M}$ compact, $\mathbf{a} \in c_0(\mathbb{Z}_{\geq 0}; \mathbb{R}_{>0})$, describing the C^ω-topology as in Sect. 2.2.6.

Definition 3.7 (Real analytic Carathéodory section). Let $\pi: \mathsf{E} \to \mathsf{M}$ be a real analytic vector bundle and let $\mathbb{T} \subseteq \mathbb{R}$ be an interval. A ***Carathéodory section of class*** C^ω of E is a map $\xi: \mathbb{T} \times \mathsf{M} \to \mathsf{E}$ with the following properties:

(i) $\xi(t, x) \in \mathsf{E}_x$ for each $(t, x) \in \mathbb{T} \times \mathsf{M}$;

(ii) for each $t \in \mathbb{T}$, the map $\xi_t: \mathsf{M} \to \mathsf{E}$ defined by $\xi_t(x)$ is of class C^ω;
(iii) for each $x \in \mathsf{M}$, the map $\xi^x: \mathbb{T} \to \mathsf{E}$ defined by $\xi^x(t) = \xi(t, x)$ is Lebesgue measurable.

We shall call $\mathbb{T}$ the ***time-domain*** for the section. By $\mathrm{CF}\Gamma^\omega(\mathbb{T}; \mathsf{E})$ we denote the set of Carathéodory sections of class C^ω of E. ∘

Now we turn to placing restrictions on the time-dependence to allow us to do useful things.

Definition 3.8 (Locally integrally C^ω-bounded and locally essentially C^ω-bounded sections). Let $\pi: \mathsf{E} \to \mathsf{M}$ be a real analytic vector bundle and let $\mathbb{T} \subseteq \mathbb{R}$ be an interval. A Carathéodory section $\xi: \mathbb{T} \times \mathsf{M} \to \mathsf{E}$ of class C^ω is

(i) ***locally integrally C^ω-bounded*** if, for every compact set $K \subseteq \mathsf{M}$ and every $\mathbf{a} \in c_0(\mathbb{Z}_{\geq 0}; \mathbb{R}_{>0})$, there exists $g \in \mathrm{L}^1_{\mathrm{loc}}(\mathbb{T}; \mathbb{R}_{\geq 0})$ such that

$$a_0 a_1 \cdots a_m \| j_m \xi_t(x) \|_{\overline{\mathbb{G}}_m} \leq g(t), \qquad (t, x) \in \mathbb{T} \times K,\ m \in \mathbb{Z}_{\geq 0},$$

and is

(ii) ***locally essentially C^ω-bounded*** if, for every compact set $K \subseteq \mathsf{M}$ and every $\mathbf{a} \in c_0(\mathbb{Z}_{\geq 0}; \mathbb{R}_{>0})$, there exists $g \in \mathrm{L}^\infty_{\mathrm{loc}}(\mathbb{T}; \mathbb{R}_{\geq 0})$ such that

$$a_0 a_1 \cdots a_m \| j_m \xi_t(x) \|_{\overline{\mathbb{G}}_m} \leq g(t), \qquad (t, x) \in \mathbb{T} \times K,\ m \in \mathbb{Z}_{\geq 0}.$$

The set of locally integrally C^ω-bounded sections of E with time-domain $\mathbb{T}$ is denoted by $\mathrm{LI}\Gamma^\omega(\mathbb{T}, \mathsf{E})$ and the set of locally essentially C^ω-bounded sections of E with time-domain $\mathbb{T}$ is denoted by $\mathrm{LB}\Gamma^\omega(\mathbb{T}; \mathsf{E})$. ∘

The following result will often be useful in practice when verifying whether a time-varying vector field is in $\mathrm{LI}\Gamma^\omega(\mathbb{T}; \mathsf{TM})$. While the result is one that is "practical", its proof relies on a comprehensive understanding of the real analytic topology.

Theorem 3.9 (Real analytic time-varying vector fields as restrictions of holomorphic time-varying vector fields). *Let $\pi: \mathsf{E} \to \mathsf{M}$ be a real analytic vector bundle with complexification*[1] *$\overline{\pi}: \overline{\mathsf{E}} \to \overline{\mathsf{M}}$, and let $\mathbb{T}$ be a time-domain. For a map $\xi: \mathbb{T} \times \mathsf{M} \to \mathsf{E}$ satisfying $\xi(t, x) \in \mathsf{E}_x$ for every $(t, x) \in \mathbb{T} \times \mathsf{M}$, the following statements hold:*

(i) if $\xi \in \mathrm{LI}\Gamma^\omega(\mathbb{T}; \mathsf{E})$, then, for each $(t_0, x_0) \in \mathbb{T} \times \mathsf{M}$ and each bounded subinterval $\mathbb{T}' \subseteq \mathbb{T}$ containing t_0, there exist a neighbourhood $\overline{\mathcal{U}}$ of x_0 in $\overline{\mathsf{M}}$ and $\overline{\xi} \in \Gamma^{\mathrm{hol}}(\mathbb{T}'; \overline{\mathsf{E}}|\overline{\mathcal{U}})$ such that $\overline{\xi}(t, x) = \xi(t, x)$ for each $t \in \mathbb{T}'$ and $x \in \overline{\mathcal{U}} \cap \mathsf{M}$;
(ii) if, for each $x_0 \in \mathsf{M}$, there exist a neighbourhood $\overline{\mathcal{U}}$ of x_0 in $\overline{\mathsf{M}}$ and $\overline{\xi} \in \Gamma^{\mathrm{hol}}(\mathbb{T}; \overline{\mathsf{E}}|\overline{\mathcal{U}})$ such that $\overline{\xi}(t, x) = \xi(t, x)$ for each $t \in \mathbb{T}$ and $x \in \overline{\mathcal{U}} \cap \mathsf{M}$, then $\xi \in \mathrm{LI}\Gamma^\omega(\mathbb{T}; \mathsf{E})$.

Proof: This is proved by Jafarpour and Lewis [7, Theorem 6.25]. □

[1] Such complexifications exist, as pointed out in [7, Sect. 5.1.1].

3.1.5 Topological Characterisations of Spaces of Time-Varying Vector Fields

The topological characterisation we give in this section relies on notions of measurability, integrability, and boundedness in the locally convex spaces $\Gamma^{\nu}(\mathsf{TM})$. Let us review this quickly for an arbitrary locally convex space V, referring to references for details.

1. A function $\gamma\colon \mathbb{T} \to \mathsf{V}$ is ***measurable*** if $\gamma^{-1}(\mathcal{O})$ is Lebesgue measurable for every open $\mathcal{O} \subseteq \mathsf{V}$.
2. It is possible to describe a notion of integral, called the ***Bochner integral***, for a function $\gamma\colon \mathbb{T} \to \mathsf{V}$ that closely resembles the usual construction of the Lebesgue integral. We refer to [7] for a sketch of the construction, and to the references cited there for details; the note [4] is particularly useful. A curve $\gamma\colon \mathbb{T} \to \mathsf{V}$ is ***Bochner integrable*** if its Bochner integral exists and is ***locally Bochner integrable*** if the Bochner integral of $\gamma|\mathbb{T}'$ exists for any compact subinterval $\mathbb{T}' \subseteq \mathbb{T}$.
3. Finally, a subset $\mathcal{B} \subseteq \mathsf{V}$ is ***von Neumann bounded*** if $p|\mathcal{B}$ is bounded for any continuous seminorm p on V. A curve $\gamma\colon \mathbb{T} \to \mathsf{V}$ is ***essentially von Neumann bounded*** if there exists a bounded set $\mathcal{B}$ such that

$$\lambda(\{t \in \mathbb{T} \mid \gamma(t) \notin \mathcal{B}\}) = 0,$$

and is ***locally essentially von Neumann bounded*** if $\gamma|\mathbb{T}'$ is essentially von Neumann bounded for every compact subinterval $\mathbb{T}' \subseteq \mathbb{T}$.

The following test for Bochner integrability and essential boundedness is one that we shall use.

Lemma 3.10 (Test for Bochner integrability and von Neumann boundedness). *Let $m \in \mathbb{Z}_{\geq 0}$, let $m' \in \{0, \mathrm{lip}\}$, let $\nu \in \{m+m', \infty, \omega, \mathrm{hol}\}$, and let $r \in \{\infty, \omega, \mathrm{hol}\}$, as required. For a manifold M of class C^r, an interval $\mathbb{T} \subseteq \mathbb{R}$, and a curve $\gamma\colon \mathbb{T} \to \Gamma^{\nu}(\mathsf{TM})$, the following two statements are equivalent:*

(i) γ is locally Bochner integrable;
(ii) for each of the seminorms p_K from (2.2), defined according to ν, there exists $g \in \mathrm{L}^1_{\mathrm{loc}}(\mathbb{T}; \mathbb{R}_{\geq 0})$ such that $p_K \circ \gamma(t) \leq g(t)$ for every $t \in \mathbb{T}$,

and the following two statements are equivalent:

(iii) γ is locally essentially von Neumann bounded;
(iv) for each of the seminorms p_K from (2.2), defined according to ν, there exists $g \in \mathrm{L}^{\infty}_{\mathrm{loc}}(\mathbb{T}; \mathbb{R}_{\geq 0})$ such that $p_K \circ \gamma(t) \leq g(t)$ for every $t \in \mathbb{T}$.

Proof: Let us prove the equivalence of the first two statements. As indicated in [7, §6], $\Gamma^{\nu}(\mathsf{TM})$ is complete and separable for all $\nu \in \{m + m', \infty, \omega\}$. By Theorems 3.1 and 3.2 of [4], we conclude that γ is locally Bochner integrable if and only if $p \circ \gamma$ is locally integrable for every continuous seminorm p for $\Gamma^{\nu}(\mathsf{TM})$. Since the seminorms from (2.2) define the locally convex topology of $\Gamma^{\nu}(\mathsf{TM})$, it suffices to check local integrability of $p_K \circ \gamma$.

The equivalence of the last two assertions follows similarly, recalling that bounded subsets of locally convex spaces are those to which restrictions of continuous seminorms are bounded. □

With these notions, we can now characterise our classes of vector fields.

Theorem 3.11 (Topological characterisation of time-varying sections). *Let* $m \in \mathbb{Z}_{\geq 0}$, *let* $m' \in \{0, \mathrm{lip}\}$, *let* $\nu \in \{m + m', \infty, \omega, \mathrm{hol}\}$, *and let* $r \in \{\infty, \omega, \mathrm{hol}\}$, *as required. For a vector bundle* $\pi: \mathsf{E} \to \mathsf{M}$ *of class* C^r *and an interval* $\mathbb{T} \subseteq \mathbb{R}$, *let* $\xi: \mathbb{T} \times \mathsf{M} \to \mathsf{E}$ *satisfy* $\xi(t, x) \in \mathsf{E}$ *for each* $(t, x) \in \mathbb{T} \times \mathsf{M}$. *Denote by* ξ_t, $t \in \mathbb{T}$, *the map* $x \mapsto \xi(t, x)$ *and suppose that* $\xi_t \in \Gamma^\nu(\mathsf{E})$ *for every* $t \in \mathbb{T}$. *Then* ξ *is:*

(i) ***a Carathéodory section of class*** $\mathbf{C}^\nu$ *if and only if the curve* $\mathbb{T} \ni t \mapsto \xi_t \in \Gamma^\nu(\mathsf{E})$ *is measurable;*
(ii) ***locally integrally*** $\mathbf{C}^\nu$ ***-bounded*** *if and only if the curve* $\mathbb{T} \ni t \mapsto \xi_t \in \Gamma^\nu(\mathsf{E})$ *is locally Bochner integrable;*
(iii) ***locally essentially*** $\mathbf{C}^\nu$ ***-bounded*** *if and only if the curve* $\mathbb{T} \ni t \mapsto \xi_t \in \Gamma^\nu(\mathsf{E})$ *is locally essentially von Neumann bounded.*

Proof: This is proved as Theorems 6.3, 6.9, 6.14, and 6.21 by Jafarpour and Lewis [7]. □

The classes of time-varying vector fields defined above have many excellent properties. Perhaps the most compelling of these is that the dependence of the flows of these vector fields on initial condition has regularity that matches ν. To be precise, if $\nu = \infty$ then the flow depends smoothly on initial condition, if $\nu = \omega$ then the flow depends real analytically on initial condition, and if $\nu = m + \mathrm{lip}$ then the flow depends m-times continuously differentiably on initial condition. These results are proved by Jafarpour and Lewis [7]; the real analytic version of this result requires a deep understanding of the C^ω-topology.

3.1.6 Mixing Regularity Hypotheses

It is possible to mix regularity conditions for vector fields. By this we mean that one can consider vector fields whose dependence on state is more regular than their joint state/time dependence. This can be done by considering $m \in \mathbb{Z}_{\geq 0} \cup \{\infty\}$, $m' \in \{0, \mathrm{lip}\}$, $r \in \mathbb{Z}_{\geq 0} \cup \{\infty, \omega\}$, and $r' \in \{0, \mathrm{lip}\}$ satisfying $m + m' < r + r'$, and considering vector fields in

$$\mathrm{CF}\Gamma^{r+r'}(\mathbb{T}; \mathsf{TM}) \cap \mathrm{LI}\Gamma^{m+m'}(\mathbb{T}; \mathsf{TM}) \quad \text{or} \quad \mathrm{CF}\Gamma^{r+r'}(\mathbb{T}; \mathsf{TM}) \cap \mathrm{LB}\Gamma^{m+m'}(\mathbb{T}; \mathsf{TM}),$$

using the obvious convention that $\infty + \mathrm{lip} = \infty$ and $\omega + \mathrm{lip} = \omega$. This does come across as quite unnatural in our framework, and perhaps it is right that it should. Moreover, because the $\mathrm{C}^{m+m'}$-topology for $\Gamma^{r+r'}(\mathsf{TM})$ will be complete if and only if $m + m' = r + r'$, some of the results above will not translate to this mixed class of time-varying vector fields: particularly, the results on Bochner integrability require completeness. Nonetheless, this mixing of regularity assumptions is quite common in the literature. Indeed, this has *always* been done in the real analytic case, since the notions of "locally integrally C^ω-bounded" and "locally essentially C^ω-bounded" given in Definition 3.8 are given for the first time in [7].

3.2 Parameterised Vector Fields

One can think of a control system as a family of vector fields parameterised by control. It is the exact nature of this dependence on the parameter that we discuss in this section. Specifically, we give pointwise characterisations that are equivalent to continuity of the natural map from the parameter space into the space of sections, using the topologies from Chap. 2.

As we have been doing thus far, we shall consider sections of general vector bundles rather than vector fields to simplify the notation.

3.2.1 The Smooth Case

We begin by discussing parameter dependent smooth sections. Throughout this section we will work with a smooth vector bundle $\pi : \mathsf{E} \to \mathsf{M}$ with a linear connection ∇^0 on E, an affine connection ∇ on M, a fibre metric $\mathbb{G}_0$ on E, and a Riemannian metric $\mathbb{G}$ on M. These define the fibre metrics $\|\cdot\|_{\overline{\mathbb{G}}_m}$ and the seminorms $p^\infty_{K,m}$, $K \subseteq \mathsf{M}$ compact, $m \in \mathbb{Z}_{\geq 0}$, on $\Gamma^\infty(\mathsf{E})$ as in Sects. 2.2.1 and 2.2.2.

Definition 3.12 **(Sections of parameterised class C^∞).** Let $\pi : \mathsf{E} \to \mathsf{M}$ be a smooth vector bundle and let $\mathcal{P}$ be a topological space. A map $\xi : \mathsf{M} \times \mathcal{P} \to \mathsf{E}$ such that $\xi(x, p) \in \mathsf{E}_x$ for every $(x, p) \in \mathsf{M} \times \mathcal{P}$

(i) is a ***separately parameterised section of class*** C^∞ if

  (a) for each $x \in \mathsf{M}$, the map $\xi_x : \mathcal{P} \to \mathsf{E}$ defined by $\xi_x(p) = \xi(x, p)$ is continuous and

  (b) for each $p \in \mathcal{P}$, the map $\xi^p : \mathsf{M} \to \mathsf{E}$ defined by $\xi^p(x) = \xi(x, p)$ is of class C^∞, and

(ii) is a ***jointly parameterised section of class*** C^∞ if it is a separately parameterised section of class C^∞ and if the map $(x, p) \mapsto j_m\xi^p(x)$ is continuous for every $m \in \mathbb{Z}_{\geq 0}$.

By $\mathrm{SP}\Gamma^\infty(\mathcal{P}; \mathsf{E})$ we denote the set of separately parameterised sections of E of class C^∞ and by $\mathrm{JP}\Gamma^\infty(\mathcal{P}; \mathsf{E})$ we denote the set of jointly parameterised sections of E of class C^∞. ○

3.2.2 The Finitely Differentiable or Lipschitz Case

The preceding development in the smooth case is easily extended to the finitely differentiable and Lipschitz cases, and we quickly give the definitions here. In this section, when considering the Lipschitz case, we assume that ∇ is the Levi-Civita connection associated to $\mathbb{G}$ and we assume that ∇^0 is $\mathbb{G}_0$-orthogonal.

Definition 3.13 (Sections of parameterised class $C^{m+m'}$). Let $\pi: \mathsf{E} \to \mathsf{M}$ be a smooth vector bundle and let $\mathcal{P}$ be a topological space. A map $\xi: \mathsf{M} \times \mathcal{P} \to \mathsf{E}$ such that $\xi(x, p) \in \mathsf{E}_x$ for every $(x, p) \in \mathsf{M} \times \mathcal{P}$

(i) is a ***separately parameterised section of class*** $C^{m+m'}$ if

  (a) for each $x \in \mathsf{M}$, the map $\xi_x: \mathcal{P} \to \mathsf{E}$ defined by $\xi_x(p) = \xi(x, p)$ is continuous and

  (b) for each $p \in \mathcal{P}$, the map $\xi^p: \mathsf{M} \to \mathsf{E}$ defined by $\xi^p(x) = \xi(x, p)$ is of class $C^{m+m'}$, and

(ii) is a ***jointly parameterised section of class*** $C^{m+m'}$ if it is a separately parameterised section of class $C^{m+m'}$ and

  (a) $m' = 0$: the map $(x, p) \mapsto j_m\xi^p(x)$ is continuous;

  (b) $m' = \mathrm{lip}$: the map $(x, p) \mapsto j_m\xi^p(x)$ is continuous and, for each $(x_0, p_0) \in \mathsf{M} \times \mathcal{P}$ and each $\varepsilon \in \mathbb{R}_{>0}$, there exist a neighbourhood $\mathcal{U} \subseteq \mathsf{M}$ of x_0 and a neighbourhood $\mathcal{O} \subseteq \mathcal{P}$ of p_0 such that

$$j_m\xi(\mathcal{U} \times \mathcal{O}) \subseteq \{j_m\eta(x) \in \mathsf{J}^m\mathsf{E} \mid \mathrm{dil}\,(j_m\eta - j_m\xi^{p_0})(x) < \varepsilon\},$$

  where, of course, $j_m\xi(x, p) = j_m\xi^p(x)$.

By $\mathrm{SP}\Gamma^{m+m'}(\mathcal{P}; \mathsf{E})$ we denote the set of separately parameterised sections of E of class $C^{m+m'}$ and by $\mathrm{JP}\Gamma^{m+m'}(\mathcal{P}; \mathsf{E})$ we denote the set of jointly parameterised sections of E of class $C^{m+m'}$. ∘

3.2.3 The Holomorphic Case

As with time-varying vector fields, we are not really interested, per se, in holomorphic control systems, and in fact we will not even define the notion. However, it is possible, and possibly sometimes easier, to verify that a control system satisfies our rather technical criterion of being a "real analytic control system" by verifying that it possesses an holomorphic extension. Thus, in this section, we present the required holomorphic definitions. We will consider an holomorphic vector bundle $\pi: \mathsf{E} \to \mathsf{M}$ with an Hermitian fibre metric $\mathbb{G}$. This defines the seminorms p_K^{hol}, $K \subseteq \mathsf{M}$ compact, describing the C^{hol}-topology for $\Gamma^{\mathrm{hol}}(\mathsf{E})$ as in Sect. 2.2.5.

Definition 3.14 (Sections of parameterised class C^{hol}). Let $\pi\colon E \to M$ be an holomorphic vector bundle and let $\mathcal{P}$ be a topological space. A map $\xi\colon M \times \mathcal{P} \to E$ such that $\xi(z, p) \in E_z$ for every $(z, p) \in M \times \mathcal{P}$

(i) is a ***separately parameterised section of class*** C^{hol} if

(a) for each $z \in M$, the map $\xi_z\colon \mathcal{P} \to E$ defined by $\xi_z(p) = \xi(z, p)$ is continuous and

(b) for each $p \in \mathcal{P}$, the map $\xi^p\colon M \to E$ defined by $\xi^p(z) = \xi(z, p)$ is of class C^{hol},

and

(ii) is a ***jointly parameterised section of class*** C^{hol} if it is a separately parameterised section of class C^{hol} and if the map $(z, p) \mapsto \xi^p(z)$ is continuous.

By $SP\Gamma^{hol}(\mathcal{P}; E)$ we denote the set of separately parameterised sections of E of class C^{hol} and by $JP\Gamma^{hol}(\mathcal{P}; E)$ we denote the set of jointly parameterised sections of E of class C^{hol}. ∘

3.2.4 The Real Analytic Case

Now we repeat the procedure above for real analytic sections. We thus will consider a real analytic vector bundle $\pi\colon E \to M$ with ∇^0 a real analytic linear connection on E, ∇ a real analytic affine connection on M, $\mathbb{G}_0$ a real analytic fibre metric on E, and $\mathbb{G}$ a real analytic Riemannian metric on M. This defines the seminorms $p^{\omega}_{K,\mathbf{a}}$, $K \subseteq M$ compact, $\mathbf{a} \in c_0(\mathbb{Z}_{\geq 0}; \mathbb{R}_{>0})$, describing the C^{ω}-topology as in Sect. 2.2.6.

Definition 3.15 (Sections of parameterised class C^{ω}). Let $\pi\colon E \to M$ be a real analytic vector bundle and let $\mathcal{P}$ be a topological space. A map $\xi\colon M \times \mathcal{P} \to E$ such that $\xi(x, p) \in E_x$ for every $(x, p) \in M \times \mathcal{P}$

(i) is a ***separately parameterised section of class*** C^{ω} if

(a) for each $x \in M$, the map $\xi_x\colon \mathcal{P} \to E$ defined by $\xi_x(p) = \xi(x, p)$ is continuous and

(b) for each $p \in \mathcal{P}$, the map $\xi^p\colon M \to E$ defined by $\xi^p(x) = \xi(x, p)$ is of class C^{ω},

and

(ii) is a ***jointly parameterised section of class*** C^{ω} if it is a separately parameterised section of class C^{∞} and if, for each $(x_0, p_0) \in M \times \mathcal{P}$, for each $\mathbf{a} \in c_0(\mathbb{Z}_{\geq 0}, \mathbb{R}_{>0})$, and for each $\varepsilon \in \mathbb{R}_{>0}$, there exist a neighbourhood $\mathcal{U} \subseteq M$ of x_0 and a neighbourhood $\mathcal{O} \subseteq \mathcal{P}$ of p_0 such that

$$j_m\xi(\mathcal{U} \times \mathcal{O}) \subseteq \{j_m\eta(x) \in J^m E \mid a_0 a_1 \cdots a_m \|j_m\eta(x) - j_m\xi^{p_0}(x)\|_{\overline{\mathbb{G}}_m} < \varepsilon\}$$

for every $m \in \mathbb{Z}_{\geq 0}$, where, of course, $j_m\xi(x, p) = j_m\xi^p(x)$.

By $\mathrm{SP}\Gamma^{\omega}(\mathcal{P};\mathsf{E})$ we denote the set of separately parameterised sections of E of class C^{ω} and by $\mathrm{JP}\Gamma^{\omega}(\mathcal{P};\mathsf{E})$ we denote the set of jointly parameterised sections of E of class C^{ω}. ○

Remark 3.16 (Jointly parameterised sections of class C^{ω}*).* The condition that $\xi \in \mathrm{JP}\Gamma^{\infty}(\mathcal{P};\mathsf{E})$ can be restated like this: for each $(x_0, p_0) \in \mathsf{M} \times \mathcal{P}$, for each $m \in \mathbb{Z}_{\geq 0}$, and for each $\varepsilon \in \mathbb{R}_{>0}$, there exist a neighbourhood $\mathcal{U} \subseteq \mathsf{M}$ of x_0 and a neighbourhood $\mathcal{O} \subseteq \mathcal{P}$ of p_0 such that

$$j_m\xi(\mathcal{U} \times \mathcal{O}) \subseteq \{j_m\eta(x) \in \mathsf{J}^m\mathsf{E} \mid \|j_m\eta(x) - j_m\xi^{p_0}(x)\|_{\overline{\mathbb{G}}_m} < \varepsilon\};$$

that this is so is, more or less, the idea of the proof of Theorem 3.18 below in the smooth case. Phrased this way, one sees clearly the grammatical similarity between the smooth and real analytic definitions. Indeed, the grammatical transformation from the smooth to the real analytic definition is, *put a factor of* $a_0 a_1 \cdots a_m$ *before the norm, precede the condition with "for every* $\boldsymbol{a} \in c_0(\mathbb{Z}_{\geq 0};\mathbb{R}_{>0})$", *and move the "for every* $m \in \mathbb{Z}_{\geq 0}$" *from before the condition to after.* This was also seen in the definitions of locally integrally bounded and locally essentially bounded sections in Sect. 3.1. Indeed, the grammatical similarity is often encountered when using our locally convex topologies, and contributes to the unification of the analysis of the varying degrees of regularity. ○

One can wonder about the relationship between sections of jointly parameterised class C^{ω} and sections that are real restrictions of sections of jointly parameterised class $\mathrm{C}^{\mathrm{hol}}$. As with Theorem 3.9 above, this is a "practical" theorem with a deep and difficult proof.

Theorem 3.17 (Jointly parameterised real analytic sections as restrictions of jointly parameterised holomorphic sections) . *Let* $\pi\colon \mathsf{E} \to \mathsf{M}$ *be a real analytic vector bundle with complexification*[2] $\overline{\pi}\colon \overline{\mathsf{E}} \to \overline{\mathsf{M}}$ *and let* $\mathcal{P}$ *be a topological space. For a map* $\xi\colon \mathsf{M} \times \mathcal{P} \to \mathsf{E}$ *satisfying* $\xi(x,p) \in \mathsf{E}_x$ *for all* $(x,p) \in \mathsf{M} \times \mathcal{P}$, *the following statements hold:*

(i) if $\xi \in \mathrm{JP}\Gamma^{\omega}(\mathcal{P};\mathsf{E})$ *and if* $\mathcal{P}$ *is locally compact and Hausdorff, then, for each* $(x_0, p_0) \in \mathsf{M} \times \mathcal{P}$, *there exist a neighbourhood* $\overline{\mathcal{U}} \subseteq \overline{\mathsf{M}}$ *of* x_0, *a neighbourhood* $\mathcal{O} \subseteq \mathcal{P}$ *of* p_0, *and* $\overline{\xi} \in \mathrm{JP}\Gamma^{\mathrm{hol}}(\mathcal{O};\overline{\mathsf{E}}|\overline{\mathcal{U}})$ *such that* $\xi(x,p) = \overline{\xi}(x,p)$ *for all* $(x,p) \in (\mathsf{M} \cap \overline{\mathcal{U}}) \times \mathcal{O}$;
(ii) if there exists a section $\overline{\xi} \in \mathrm{JP}\Gamma^{\mathrm{hol}}(\mathcal{P};\overline{\mathsf{E}})$ *such that* $\xi(x,p) = \overline{\xi}(x,p)$ *for every* $(x,p) \in \mathsf{M} \times \mathcal{P}$, *then* $\xi \in \mathrm{JP}\Gamma^{\omega}(\mathcal{P};\mathsf{E})$.

Proof: This is proved by Jafarpour and Lewis [6, Theorem 4.10]. □

[2] Such complexifications exist, as pointed out in [7, Sect. 5.1.1].

3.2.5 Topological Characterisations of Parameterised Vector Fields

As with time-varying vector fields, it is possible to characterise parameterised vector fields using the locally convex topologies developed in Chap. 2. This is done relatively easily here, since we only rely on continuity properties, not on the more complicated notions of measurability and integrability we used for time-varying vector fields.

Theorem 3.18 (Topological characterisation of parameterised vector fields). *Let $m \in \mathbb{Z}_{\geq 0}$, let $m' \in \{0, \text{lip}\}$, let $\nu \in \{m + m', \infty, \omega, \text{hol}\}$, and let $r \in \{\infty, \omega, \text{hol}\}$, as required. For a vector bundle $\pi: \mathsf{E} \to \mathsf{M}$ of class C^r and a topological space $\mathcal{P}$, let $\xi: \mathsf{M} \times \mathcal{P} \to \mathsf{E}$ satisfy $\xi(x, p) \in \mathsf{E}_x$ for each $(x, p) \in \mathsf{M} \times \mathcal{P}$. Then $\xi \in \mathrm{JP}\Gamma^\nu(\mathcal{P}; \mathsf{E})$ if and only if the map $p \mapsto \xi^p \in \Gamma^\nu(\mathsf{E})$ is continuous, where $\Gamma^\nu(\mathsf{E})$ has the C^ν-topology.*

Proof: This is proved as Propositions 4.2, 4.4, 4.6, and 4.9 by Jafarpour and Lewis [6]. □

3.2.6 Mixing Regularity Hypotheses

It is possible to consider parameterised sections with mixed regularity hypotheses. Indeed, the conditions of Definitions 3.12, 3.13, and 3.15 are joint on state and parameter. Thus we may consider the following situation. Let $m \in \mathbb{Z}_{\geq 0} \cup \{\infty\}$, $m' \in \{0, \text{lip}\}$, $r \in \mathbb{Z}_{\geq 0} \cup \{\infty, \omega\}$, and $r' \in \{0, \text{lip}\}$. If $r + r' \geq m + m'$ (with the obvious convention that $\infty + \text{lip} = \infty$ and $\omega + \text{lip} = \omega$), we may then consider a parameterised section in

$$\mathrm{SP}\Gamma^{r+r'}(\mathcal{P}; \mathsf{E}) \cap \mathrm{JP}\Gamma^{m+m'}(\mathcal{P}; \mathsf{E})$$

As with time-varying vector fields, there is nothing wrong with this—indeed this is often done—as long as one remembers what is true and what is not in the case when $r + r' > m + m'$.

3.3 Control Systems

In this section we shall present a class of control systems of the "ordinary" sort. These systems, while of a standard form, are defined in such a way that the appropriate topology for the space of vector fields is carefully accounted for. In Sect. 3.3.3 we also briefly discuss differential inclusions.

3.3.1 Control Systems with Locally Essentially Bounded Controls

With the notions of parameterised sections from the preceding section, we readily define what we mean by a control system.

Definition 3.19 (Control system). Let $m \in \mathbb{Z}_{\geq 0}$ and $m' \in \{0, \text{lip}\}$, let $\nu \in \{m + m', \infty, \omega\}$, and let $r \in \{\infty, \omega\}$, as required. A $\mathbf{C}^{\nu}$***-control system*** is a triple $\Sigma = (\mathsf{M}, F, \mathcal{C})$, where

(i) M is a C^r-manifold whose elements are called ***states***,
(ii) $\mathcal{C}$ is a topological space called the ***control set***, and
(iii) $F \in \mathrm{JP}\Gamma^{\nu}(\mathcal{C}; \mathsf{TM})$. ∘

The governing equations for a control system are

$$\xi'(t) = F(\xi(t), \mu(t)),$$

for suitable functions $t \mapsto \mu(t) \in \mathcal{C}$ and $t \mapsto \xi(t) \in \mathsf{M}$. To ensure that these equations make sense, the differential equation should be shown to have the properties needed for existence and uniqueness of solutions, as well as appropriate dependence on initial conditions. We do this by allowing the controls for the system to be as general as reasonable.

Proposition 3.20 (Property of control system when the control is specified). *Let $m \in \mathbb{Z}_{\geq 0}$ and $m' \in \{0, \text{lip}\}$, let $\nu \in \{m + m', \infty, \omega\}$, and let $r \in \{\infty, \omega\}$, as required. Let $\Sigma = (\mathsf{M}, F, \mathcal{C})$ be a C^{ν}-control system. If $\mu \in L^{\text{cpt}}_{\text{loc}}(\mathbb{T}; \mathcal{C})$ then $F^{\mu} \in \mathrm{LB}\Gamma^{\nu}(\mathbb{T}, \mathsf{TM})$, where $F^{\mu}\colon \mathbb{T} \times \mathsf{M} \to \mathsf{TM}$ is defined by $F^{\mu}(t, x) = F(x, \mu(t))$.*

Proof: We refer to [6, Proposition 5.2]. □

The notion of a trajectory is, of course, well known. However, we make the definitions clear for future reference.

Definition 3.21 (Trajectory for control system). Let $m \in \mathbb{Z}_{\geq 0}$ and $m' \in \{0, \text{lip}\}$, let $\nu \in \{m + m', \infty, \omega\}$, and let $r \in \{\infty, \omega\}$, as required. Let $\Sigma = (\mathsf{M}, F, \mathcal{C})$ be a C^{ν}-control system. For an interval $\mathbb{T} \subseteq \mathbb{R}$, a $\mathbb{T}$***-trajectory*** is a locally absolutely continuous curve $\xi\colon \mathbb{T} \to \mathsf{M}$ for which there exists $\mu \in L^{\text{cpt}}_{\text{loc}}(\mathbb{T}; \mathcal{C})$ such that

$$\xi'(t) = F(\xi(t), \mu(t)), \qquad \text{a.e. } t \in \mathbb{T}.$$

The set of $\mathbb{T}$-trajectories we denote by $\mathrm{Traj}(\mathbb{T}, \Sigma)$. If $\mathcal{U}$ is open, we denote by $\mathrm{Traj}(\mathbb{T}, \mathcal{U}, \Sigma)$ those trajectories taking values in $\mathcal{U}$.[3] ∘

3.3.2 Control Systems with Locally Integrable Controls

In this section we specialise the discussion from the preceding section in one direction, while generalising it in another. To be precise, we now consider the case where

[3] This is not a common notion in this context, and our introduction of this is for the convenience of making comparisons when we talk about tautological control systems in Chap. 5; see Theorems 5.27 and 5.29.

our control set $\mathcal{C}$ is a subset of a locally convex topological vector space, and the system structure is such that the notion of integrability is preserved (in a way that will be made clear in Proposition 3.24 below).

Definition 3.22 (Sublinear control system). Let $m \in \mathbb{Z}_{\geq 0}$ and $m' \in \{0, \text{lip}\}$, let $\nu \in \{m + m', \infty, \omega\}$, and let $r \in \{\infty, \omega\}$, as required. A C^ν ***-sublinear control system*** is a triple $\Sigma = (\mathsf{M}, F, \mathcal{C})$, where

(i) M is a C^r-manifold whose elements are called ***states***,
(ii) $\mathcal{C}$ is a subset of a locally convex topological vector space V, $\mathcal{C}$ being called the ***control set***, and
(iii) $F: \mathsf{M} \times \mathcal{C} \to \mathsf{TM}$ has the following property: for every continuous seminorm p for $\Gamma^\nu(\mathsf{TM})$, there exists a continuous seminorm q for V such that

$$p(F^{u_1} - F^{u_2}) \leq q(u_1 - u_2), \qquad u_1, u_2 \in \mathcal{C}. \qquad \circ$$

Note that, by Theorem 3.18, the sublinearity condition (iii) implies that a C^ν-sublinear control system is a C^ν-control system.

Let us demonstrate a class of sublinear control systems in which we will be particularly interested.

Example 3.23 (Control-linear systems and control-affine systems). The class of sublinear control systems we consider seems quite particular, but will turn out to be extremely general in our framework. We let $m \in \mathbb{Z}_{\geq 0}$ and $m' \in \{0, \text{lip}\}$, let $\nu \in \{m + m', \infty, \omega\}$, and let $r \in \{\infty, \omega\}$, as required. Let V be a locally convex topological vector space and let $\mathcal{C} \subseteq \mathsf{V}$. We suppose that we have a continuous linear map $\Lambda \in \mathrm{L}(\mathsf{V}; \Gamma^\nu(\mathsf{TM}))$ and we correspondingly define $F_\Lambda: \mathsf{M} \times \mathcal{C} \to \mathsf{TM}$ by $F_\Lambda(x, u) = \Lambda(u)(x)$. Continuity of Λ immediately gives that the control system $(\mathsf{M}, F_\Lambda, \mathcal{C})$ is sublinear, and we shall call a system such as this a $\mathbf{C}^\nu$***-control-linear system***.

Note that we can regard a control-affine system as a control-linear system as follows. For a control-affine system with $\mathcal{C} \subseteq \mathbb{R}^k$ and with

$$F(x, \mathbf{u}) = f_0(x) + \sum_{a=1}^{k} u^a f_a(x),$$

we let $\mathsf{V} = \mathbb{R}^{k+1} \simeq \mathbb{R} \oplus \mathbb{R}^k$ and take

$$\mathcal{C}' = \{(u^0, \mathbf{u}) \in \mathbb{R} \oplus \mathbb{R}^k \mid u^0 = 1, \ \mathbf{u} \in \mathcal{C}\}, \quad \Lambda(u^0, \mathbf{u}) = \sum_{a=0}^{k} u^a f_a.$$

Clearly we have $F(x, \mathbf{u}) = F_\Lambda(x, (1, \mathbf{u}))$ for every $\mathbf{u} \in \mathcal{C}$. Since linear maps from finite-dimensional locally convex spaces are continuous [5, Proposition 2.10.2], we conclude that control-affine systems are control-linear systems. Thus they are also control systems as per Definition 3.19. $\circ$

One may want to regard the generalisation from the case where the control set is a subset of $\mathbb{R}^k$ to being a subset of a locally convex topological vector space to be mere fancy generalisation, but this is, actually, far from being the case. Indeed, this observation is the foundation for the connections we make in Chap. 5 between "ordinary" control systems and tautological control systems. We also have a version of Proposition 3.20 for sublinear control systems.

Proposition 3.24 (Property of sublinear control system when the control is specified). *Let $m \in \mathbb{Z}_{\geq 0}$ and $m' \in \{0, \text{lip}\}$, let $\nu \in \{m + m', \infty, \omega\}$, and let $r \in \{\infty, \omega\}$, as required. Let $\Sigma = (\mathsf{M}, F, \mathcal{C})$ be a C^ν-sublinear control system for which $\mathcal{C}$ is a subset of a locally convex topological vector space V. If $\mu \in \mathrm{L}^1_{\text{loc}}(\mathbb{T}; \mathcal{C})$, then $F^\mu \in \mathrm{LI}\Gamma^\nu(\mathbb{T}; \mathsf{TM})$, where $F^\mu : \mathbb{T} \times \mathsf{M} \to \mathsf{TM}$ is defined by $F^\mu(t, x) = F(x, \mu(t))$.*

Proof: We refer to [6, Proposition 5.6]. □

There is also a version of the notion of trajectory that is applicable to the case when the control set is a subset of a locally convex topological space.

Definition 3.25 (Trajectory for sublinear control system). Let $m \in \mathbb{Z}_{\geq 0}$ and $m' \in \{0, \text{lip}\}$, let $\nu \in \{m+m', \infty, \omega\}$, and let $r \in \{\infty, \omega\}$, as required. Let $\Sigma = (\mathsf{M}, F, \mathcal{C})$ be a C^ν-sublinear control system. For an interval $\mathbb{T} \subseteq \mathbb{R}$, a ***$\mathbb{T}$-trajectory*** is a locally absolutely continuous curve $\xi : \mathbb{T} \to \mathsf{M}$ for which there exists $\mu \in \mathrm{L}^1_{\text{loc}}(\mathbb{T}; \mathcal{C})$ such that

$$\xi'(t) = F(\xi(t), \mu(t)), \qquad \text{a.e. } t \in \mathbb{T}.$$

The set of $\mathbb{T}$-trajectories we denote by $\text{Traj}(\mathbb{T}, \Sigma)$. If $\mathcal{U}$ is open, we denote by $\text{Traj}(\mathbb{T}, \mathcal{U}, \Sigma)$ those trajectories taking values in $\mathcal{U}$. ○

3.3.3 Differential Inclusions

We briefly mentioned differential inclusions in Sect. 1.1.4, but now let us define them properly and give a few attributes of, and constructions for, differential inclusions of which we shall subsequently make use.

First the definition.

Definition 3.26 (Differential inclusion, trajectory). For a smooth manifold M, a ***differential inclusion*** on M is a set-valued map $\mathscr{X} : \mathsf{M} \twoheadrightarrow \mathsf{TM}$ with nonempty values for which $\mathscr{X}(x) \subseteq \mathsf{T}_x\mathsf{M}$. A ***trajectory*** for a differential inclusion $\mathscr{X}$ is a locally absolutely continuous curve $\xi : \mathbb{T} \to \mathsf{M}$ defined on an interval $\mathbb{T} \subseteq \mathbb{R}$ for which $\xi'(t) \in \mathscr{X}(\xi(t))$ for almost every $t \in \mathbb{T}$. If $\mathbb{T} \subseteq \mathbb{R}$ is an interval and if $\mathcal{U} \subseteq \mathsf{M}$ is open, by $\text{Traj}(\mathbb{T}, \mathcal{U}, \mathscr{X})$ we denote the trajectories of $\mathscr{X}$ defined on $\mathbb{T}$ and taking values in $\mathcal{U}$. ○

Of course, differential inclusions will generally not have trajectories, and to ensure that they do various hypotheses can be made. Two common attributes of differential inclusions in this vein are the following.

Definition 3.27 (Lower and upper semicontinuity of differential inclusions). A differential inclusion $\mathscr{X}$ on a smooth manifold M is:

(i) ***lower semicontinuous*** at $x_0 \in \mathsf{M}$ if, for any $v_0 \in \mathscr{X}(x_0)$ and any neighbourhood $\mathcal{V} \subseteq \mathsf{TM}$ of v_0, there exists a neighbourhood $\mathcal{U} \subseteq \mathsf{M}$ of x_0 such that $\mathscr{X}(x) \cap \mathcal{V} \neq \emptyset$ for every $x \in \mathcal{U}$;
(ii) ***lower semicontinuous*** if it is lower semicontinuous at every $x \in \mathsf{M}$;
(iii) ***upper semicontinuous*** at $x_0 \in \mathsf{M}$ if, for every open set $\mathsf{TM} \supset \mathcal{V} \supset \mathscr{X}(x_0)$, there exists a neighbourhood $\mathcal{U} \subseteq \mathsf{M}$ of x_0 such that $\mathscr{X}(\mathcal{U}) \subseteq \mathcal{V}$;
(iv) ***upper semicontinuous*** if it is upper semicontinuous at each $x \in \mathsf{M}$;
(v) ***continuous*** at $x_0 \in \mathsf{M}$ if it is both lower and upper semicontinuous at x_0;
(vi) ***continuous*** if it is both lower and upper semicontinuous. ∘

Other useful properties of differential inclusions are the following.

Definition 3.28 (Closed-valued, compact-valued, convex-valued differential inclusions). A differential inclusion $\mathscr{X}$ on a smooth manifold M is:

(i) ***closed-valued*** (resp. ***compact-valued***, ***convex-valued***) at $x \in \mathsf{M}$ if $\mathscr{X}(x)$ is closed (resp., compact, convex);
(ii) ***closed-valued*** (resp. ***compact-valued***, ***convex-valued***) if $\mathscr{X}(x)$ is closed (resp., compact, convex) for every $x \in \mathsf{M}$. ∘

Some standard hypotheses for existence of trajectories are then:

1. $\mathscr{X}$ is lower semicontinuous with closed and convex values [3, Theorem 2.1.1];
2. $\mathscr{X}$ is upper semicontinuous with compact and convex values [3, Theorem 2.1.4];
3. $\mathscr{X}$ is continuous with compact values [3, Theorem 2.2.1].

These are not matters with which we shall be especially concerned.

A standard operation is to take "hulls" of differential inclusions in the following manner.

Definition 3.29 (Convex hull, closure of a differential inclusion). Let $r \in \{\infty, \omega\}$, let M be a C^r-manifold, and let $\mathscr{X} : \mathsf{M} \twoheadrightarrow \mathsf{TM}$ be a differential inclusion.

(i) The ***convex hull*** of $\mathscr{X}$ is the differential inclusion $\operatorname{conv}(\mathscr{X})$ defined by

$$\operatorname{conv}(\mathscr{X})(x) = \operatorname{conv}(\mathscr{X}(x)), \qquad x \in \mathsf{M}.$$

(ii) The ***closure*** of $\mathscr{X}$ is the differential inclusion $\operatorname{cl}(\mathscr{X})$ defined by

$$\operatorname{cl}(\mathscr{X})(x) = \operatorname{cl}(\mathscr{X}(x)), \qquad x \in \mathsf{M}.$$ ∘

To close this section, let us make an observation regarding the connection between control systems and differential inclusions. Let $m \in \mathbb{Z}_{\geq 0}$ and $m' \in \{0, \mathrm{lip}\}$, let $\nu \in \{m + m', \infty, \omega\}$, and let $r \in \{\infty, \omega\}$, as required. Let $\Sigma = (\mathsf{M}, F, \mathcal{C})$ be a C^ν-control system. To this system we associate the differential inclusion $\mathscr{X}_\Sigma$ by

$$\mathscr{X}_\Sigma(x) = \{F^u(x) \mid u \in \mathcal{C}\}.$$

Since the differential inclusion $\mathscr{X}_\Sigma$ is defined by a family of vector fields, one might try to recover the vector fields F^u, $u \in \mathcal{C}$, from $\mathscr{X}_\Sigma$. The obvious way to do this is to consider

$$\Gamma^\nu(\mathscr{X}_\Sigma) \triangleq \{X \in \Gamma^\nu(\mathsf{TM}) \mid X(x) \in \mathscr{X}_\Sigma(x),\ x \in \mathsf{M}\}.$$

Clearly we have $F^u \in \Gamma^\nu(\mathscr{X}_\Sigma)$ for every $u \in \mathcal{C}$. However, $\mathscr{X}_\Sigma$ will generally contain vector fields not of the form F^u for some $u \in \mathcal{C}$. Let us give an illustration of this. Let us consider a smooth control system $(\mathsf{M}, F, \mathcal{C})$ with the following properties:

1. $\mathcal{C}$ is a disjoint union of sets $\mathcal{C}_1$ and $\mathcal{C}_2$;
2. there exist disjoint open sets $\mathcal{U}_1$ and $\mathcal{U}_2$ such that $\mathrm{supp}(F^u) \subseteq \mathcal{U}_1$ for $u \in \mathcal{C}_1$ and $\mathrm{supp}(F^u) \subseteq \mathcal{U}_2$ for $u \in \mathcal{C}_2$.

One then has that

$$\{c_1 F^{u_1} + c_1 F^{u_2} \mid u_1 \in \mathcal{C}_1,\ u_2 \in \mathcal{C}_2,\ c_1, c_2 \in \{0, 1\},\ c_1^2 + c_2^2 \neq 0\} \subseteq \Gamma^\nu(\mathscr{X}_\Sigma),$$

showing that there are more sections of $\mathscr{X}_\Sigma$ than there are control vector fields. This is very much related to presheaves and sheaves, to which we shall now turn our attention.

References

1. Agrachev AA, Gamkrelidze RV (1978) The exponential representation of flows and the chronological calculus. Math USSR-Sb 107(4):467–532
2. Agrachev AA, Sachkov Y (2004) Control theory from the geometric viewpoint, Encyclopedia of Mathematical Sciences, vol 87. Springer, Berlin
3. Aubin JP, Cellina A (1984) Differential inclusions: set-valued maps and viability theory, Grundlehren der Mathematischen Wissenschaften, vol 264. Springer, Berlin
4. Beckmann R, Deitmar A (2011) Strong vector valued integrals. ArXiv:1102.1246v1 [math.FA]. http://arxiv.org/abs/1102.1246v1
5. Horváth J (1966) Topological vector spaces and distributions, vol I. Addison Wesley, Reading, MA
6. Jafarpour S, Lewis AD (2014) Locally convex topologies and control theory. Submitted to SIAM J Control Optim
7. Jafarpour S, Lewis AD (2014) Time-varying vector fields and their flows. To appear in Springer Briefs in Mathematics
8. Sussmann HJ (1997) An introduction to the coordinate-free maximum principle. In: Jakubczyk B, Respondek W (eds) Geometry of feedback and optimal control. Dekker Marcel Dekker, New York, pp 463–557
9. Whitney H (1936) Differentiable manifolds. Ann Math 37(3):645–680

Chapter 4
Presheaves and Sheaves of Sets of Vector Fields

We choose to phrase our notion of control systems in the language of sheaf theory. This will seem completely pointless to a reader not used to thinking in this sort of language. However, we do believe there are benefits to the sheaf approach including (1) it permits natural formulations of problems that do not have a natural formulation in "ordinary" control theory, (2) sheaves are the proper framework for constructing germs of control systems which are often important in the study of local system structure, and (3) sheaf theory provides us with a natural class of mappings between systems that we use to advantage in Sect. 5.6.

4.1 Definitions and Examples

We do not even come close to discussing sheaves in any generality; we merely give the definitions we require, a few of the most elementary consequences of these definitions, and some representative (for us) examples. We refer to [3, 4, 7–9] for details.

Definition 4.1 (Presheaf of sets of vector fields). Let $m \in \mathbb{Z}_{\geq 0}$ and $m' \in \{0, \mathrm{lip}\}$, let $\nu \in \{m + m', \infty, \omega\}$, and let $r \in \{\infty, \omega\}$, as required. Let M be a manifold of class C^r. A ***presheaf of sets of*** C^ν***-vector fields*** is an assignment to each open set $\mathcal{U} \subseteq \mathsf{M}$ a subset $\mathscr{F}(\mathcal{U})$ of $\Gamma^\nu(\mathsf{T}\mathcal{U})$ with the property that, for open sets $\mathcal{U}, \mathcal{V} \subseteq \mathsf{M}$ with $\mathcal{V} \subseteq \mathcal{U}$, the map

$$\begin{aligned} r_{\mathcal{U},\mathcal{V}}\colon \mathscr{F}(\mathcal{U}) &\to \Gamma^\nu(\mathsf{T}\mathcal{V}) \\ X &\mapsto X|\mathcal{V} \end{aligned}$$

takes values in $\mathscr{F}(\mathcal{V})$. Elements of $\mathscr{F}(\mathcal{U})$ are called ***local sections*** of $\mathscr{F}$ over $\mathcal{U}$. ∘

Let us give some notation to the presheaf of sets of vector fields of which every other such presheaf is a subset.

A.D. Lewis, *Tautological Control Systems*, SpringerBriefs in Control, Automation and Robotics, DOI: 10.1007/978-3-319-08638-5_4

Example 4.2 (Presheaf of all vector fields). Let $m \in \mathbb{Z}_{\geq 0}$ and $m' \in \{0, \text{lip}\}$, let $\nu \in \{m + m', \infty, \omega\}$, and let $r \in \{\infty, \omega\}$, as required. Let M be a manifold of class C^r. The presheaf of *all* vector fields of class C^ν is denoted by $\mathscr{G}^\nu_{\mathsf{TM}}$. Thus $\mathscr{G}^\nu_{\mathsf{TM}}(\mathcal{U}) = \Gamma^\nu(\mathsf{T}\mathcal{U})$ for every open set $\mathcal{U}$. Presheaves such as this are extremely important in the "normal" applications of sheaf theory. For those with some background in these more standard applications of sheaf theory, we mention that our reasons for using the theory are not quite the usual ones. Such readers will be advised to be careful not to overlay too much of their past experience on what we do with sheaf theory here. ∘

The preceding notion of a presheaf is intuitively clear, but it does have some defects. One of these defects is that one can describe local data that does not patch together to give global data. Let us illustrate this with a few examples.

Examples 4.3 (Local definitions not globally consistent).

1. Let $m \in \mathbb{Z}_{\geq 0}$ and $m' \in \{0, \text{lip}\}$, let $\nu \in \{m + m', \infty, \omega\}$, and let $r \in \{\infty, \omega\}$, as required. Let us take a manifold M of class C^r with a Riemannian metric $\mathbb{G}$. Let us define a presheaf $\mathscr{F}_{\text{bdd}}$ by asking that

$$\mathscr{F}_{\text{bdd}}(\mathcal{U}) = \{X \in \Gamma^\nu(\mathsf{TM}) \mid \sup\{\|X(x)\|_{\mathbb{G}} \mid x \in \mathcal{U}\} < \infty\}.$$

Thus $\mathscr{F}_{\text{bdd}}$ is comprised of vector fields that are "bounded". This is a perfectly sensible requirement. However, the following phenomenon can happen if M is not compact. There can exist an open cover $(\mathcal{U}_a)_{a\in A}$ for M and local sections $X_a \in \mathscr{F}_{\text{bdd}}(\mathcal{U}_a)$ that are "compatible" in the sense that $X_a|\mathcal{U}_a \cap \mathcal{U}_b = X_b|\mathcal{U}_a \cap \mathcal{U}_b$, for each $a, b \in A$, but such that there is no globally defined section $X \in \mathscr{F}_{\text{bdd}}(\mathsf{M})$ such that $X|\mathcal{U}_a = X_a$ for every $a \in A$. We leave to the reader the easy job of coming up with a concrete instance of this.

2. We let $m \in \mathbb{Z}_{\geq 0}$ and $m' \in \{0, \text{lip}\}$, let $\nu \in \{m + m', \infty, \omega\}$, and let $r \in \{\infty, \omega\}$, as required. Let M be a manifold of class C^r and let $(\mathcal{U}_a)_{a\in A}$ be an open cover of M. For each $a \in A$ suppose that we have $X_a \in \Gamma^\nu(\mathsf{T}\mathcal{U}_a)$; this echoes the setting of the Orbit Theorem of Stefan [10] and Sussmann [11]. We define a presheaf $\mathscr{F}$ of sets of C^ν-vector fields as follows. For $\mathcal{U} \subseteq \mathsf{M}$ open, we define

$$\mathscr{F}(\mathcal{U}) = \{X \in \Gamma^\nu(\mathsf{T}\mathcal{U}) \mid \text{there exists } a \in A \text{ such that } \mathcal{U} \subseteq \mathcal{U}_a \text{ and } X = X_a|\mathcal{U}\}.$$

The verification that this is a presheaf is easily carried out. However, it will generally not have decent global patching properties. This can arise in many ways for this presheaf, and here is one. Suppose that $a, b \in A$ are such that $\mathcal{U}_a \cap \mathcal{U}_b = \emptyset$. We take $\mathcal{U} = \mathcal{U}_a \cup \mathcal{U}_b$ and consider the open cover $(\mathcal{U}_a, \mathcal{U}_b)$ for $\mathcal{U}$. Then the vector fields X_a and X_b vacuously agree on $\mathcal{U}_a \cap \mathcal{U}_b$. We then define $X \in \Gamma^\nu(\mathsf{T}\mathcal{U})$ by

$$X(x) = \begin{cases} X_a(x), & x \in \mathcal{U}_a, \\ X_b(x), & x \in \mathcal{U}_b. \end{cases}$$

Generally, however, we will not have $\mathcal{U} \subseteq \mathcal{U}_c$ for some $c \in A$, and so $X \notin \mathscr{F}(\mathcal{U})$.

3. Let $m \in \mathbb{Z}_{\geq 0}$ and $m' \in \{0, \text{lip}\}$, let $\nu \in \{m + m', \infty, \omega\}$, and let $r \in \{\infty, \omega\}$, as required. Let M be a manifold of class C^r. If $\mathscr{X} \subseteq \Gamma^\nu(\mathsf{TM})$ is any family of vector fields on M, then we can define an associated presheaf $\mathscr{F}_{\mathscr{X}}$ of sets of vector fields by

$$\mathscr{F}_{\mathscr{X}}(\mathcal{U}) = \{X|\mathcal{U} \mid X \in \mathscr{X}\}.$$

Note that $\mathscr{F}(\mathsf{M})$ is necessarily equal to $\mathscr{X}$, and so we shall typically use $\mathscr{F}(\mathsf{M})$ to denote the set of globally defined vector fields giving rise to this presheaf. A presheaf of this sort will be called ***globally generated***.

This sort of presheaf will almost never have nice "local to global" properties. Let us illustrate why this is so. Let M be a connected Hausdorff manifold. Suppose that the set of globally defined vector fields $\mathscr{F}(\mathsf{M})$ has cardinality strictly larger than 1 and has the following property: there exists a disconnected open set $\mathcal{U} \subseteq \mathsf{M}$ such that the mapping from $\mathscr{F}(\mathcal{U})$ to $\mathscr{F}(\mathsf{M})$ given by $X|\mathcal{U} \mapsto X$ is injective. This property will hold for real analytic families of vector fields, because we can take as $\mathcal{U}$ the union of a pair of disconnected open sets. However, the property will also hold for many reasonable families of smooth vector fields.

We write $\mathcal{U} = \mathcal{U}_1 \cup \mathcal{U}_2$ for disjoint open sets $\mathcal{U}_1$ and $\mathcal{U}_2$. By hypothesis, there exist vector fields $X_1, X_2 \in \mathscr{F}(\mathsf{M})$ such that $X_1|\mathcal{U} \neq X_2|\mathcal{U}$. Define local sections $X'_a \in \mathscr{F}(\mathcal{U}_a)$ by $X'_a = X_a|\mathcal{U}_a$, $a \in \{1, 2\}$. The condition

$$X'_1|\mathcal{U}_1 \cap \mathcal{U}_2 = X'_2|\mathcal{U}_1 \cap \mathcal{U}_2$$

is vacuously satisfied. But there can be no $X \in \mathscr{F}(\mathsf{M})$ such that, if $X' \in \mathscr{F}(\mathcal{U})$ is given by $X' = X|\mathcal{U}$, then $X'|\mathcal{U}_1 = X'_1$ and $X'|\mathcal{U}_2 = X'_2$.

While a globally generated presheaf is unlikely to allow patching from local to global, this can be easily redressed by undergoing a process known as "sheafification" that we will describe below. ○

The preceding examples suggest that, if one wishes to make compatible local constructions that give rise to a global construction, additional properties need to be ascribed to a presheaf of sets of vector fields. This we do as follows.

Definition 4.4 (Sheaf of sets of vector fields). Let $m \in \mathbb{Z}_{\geq 0}$ and $m' \in \{0, \text{lip}\}$, let $\nu \in \{m + m', \infty, \omega\}$, and let $r \in \{\infty, \omega\}$, as required. Let M be a manifold of class C^r. A presheaf $\mathscr{F}$ of sets of C^ν-vector fields is a ***sheaf of sets of*** $\mathbf{C}^\nu$***-vector fields*** if, for every open set $\mathcal{U} \subseteq \mathsf{M}$, for every open cover $(\mathcal{U}_a)_{a \in A}$ of $\mathcal{U}$, and for every choice of local sections $X_a \in \mathscr{F}(\mathcal{U}_a)$ satisfying $X_a|\mathcal{U}_a \cap \mathcal{U}_b = X_b|\mathcal{U}_a \cap \mathcal{U}_b$, there exists $X \in \mathscr{F}(\mathcal{U})$ such that $X|\mathcal{U}_a = X_a$ for every $a \in A$. ○

The condition in the definition is called the ***gluing condition***. Readers familiar with sheaf theory will note the absence of another condition, sometimes called the separation condition, normally placed on a presheaf in order for it to be a sheaf: it is automatically satisfied for presheaves of sets of vector fields.

Many of the presheaves that we encounter will not be sheaves, as they will be globally generated. Thus let us give some examples of sheaves, just as a point of reference.

Examples 4.5 (Sheaves of sets of vector fields).

1. Let $m \in \mathbb{Z}_{\geq 0}$ and $m' \in \{0, \mathrm{lip}\}$, let $\nu \in \{m + m', \infty, \omega\}$, and let $r \in \{\infty, \omega\}$, as required. Let M be a C^r-manifold. The presheaf $\mathscr{G}^{\nu}_{\mathsf{TM}}$ of all C^{ν}-vector fields is a sheaf. We leave the simple and standard working out of this to the reader; it will provide some facility in working with sheaf concepts for those not already having this.
2. If instead of considering bounded vector fields as in part Example 4.3–1, we consider the presheaf of vector fields satisfying a *fixed* bound, then the resulting presheaf is a sheaf. Let us be clear. Let $m \in \mathbb{Z}_{\geq 0}$ and $m' \in \{0, \mathrm{lip}\}$, let $\nu \in \{m+m', \infty, \omega\}$, and let $r \in \{\infty, \omega\}$, as required. We let M be a C^r-manifold with Riemannian metric $\mathbb{G}$ and, for $B \in \mathbb{R}_{>0}$, define a presheaf $\mathscr{F}_{\leq B}$ by
$$\mathscr{F}_{\leq B}(\mathcal{U}) = \{X \in \Gamma^{\nu}(\mathsf{T}\mathcal{U}) \mid \sup\{\|X(x)\|_{\mathbb{G}} \mid x \in \mathcal{U}\} \leq B\}.$$
The presheaf $\mathscr{F}_{\leq B}$ is a sheaf, as is easily verified. In this case, the local constraints for membership are compatible with a global one.
3. Let $m \in \mathbb{Z}_{\geq 0}$ and $m' \in \{0, \mathrm{lip}\}$, let $\nu \in \{m + m', \infty, \omega\}$, and let $r \in \{\infty, \omega\}$, as required. Let M be a C^r-manifold. Let $A \subseteq \mathsf{M}$ and define a presheaf $\mathscr{I}_A$ of sets of vector fields by
$$\mathscr{I}_A(\mathcal{U}) = \{X \in \Gamma^{\nu}(\mathsf{T}\mathcal{U}) \mid X(x) = 0,\ x \in A\}.$$
This is a sheaf (again, we leave the verification to the reader) called the ***ideal sheaf*** of A. ○

Let us now turn to localising sheaves of sets of vector fields. Let $m \in \mathbb{Z}_{\geq 0}$ and $m' \in \{0, \mathrm{lip}\}$, let $\nu \in \{m + m', \infty, \omega\}$, and let $r \in \{\infty, \omega\}$, as required. Let M be a C^r-manifold, let $A \subseteq \mathsf{M}$, and let $\mathscr{N}_A$ be the set of neighbourhoods of A in M, i.e., the open subsets of M containing A. This is a directed set in the usual way by inclusion, i.e., $\mathcal{U} \preceq \mathcal{V}$ if $\mathcal{V} \subseteq \mathcal{U}$. Let $\mathscr{F}$ be a sheaf of sets of C^{ν}-vector fields. The ***stalk*** of $\mathscr{F}$ over A is the direct limit $\mathrm{dirlim}_{\mathcal{U} \in \mathscr{N}_A} \mathscr{F}(\mathcal{U})$. Let us be less cryptic about this. Let $\mathcal{U}, \mathcal{V} \in \mathscr{N}_A$, and let $X \in \mathscr{F}(\mathcal{U})$ and $Y \in \mathscr{F}(\mathcal{V})$. We say X and Y are ***equivalent*** if there exists $\mathcal{W} \subseteq \mathcal{U} \cap \mathcal{V}$ such that $X|\mathcal{W} = Y|\mathcal{W}$. The ***germ*** of $X \in \mathscr{F}(\mathcal{U})$ for $\mathcal{U} \in \mathscr{N}_A$ is the equivalence class of X under this equivalence relation. If $\mathcal{U} \in \mathscr{N}_A$ and if $X \in \mathscr{F}(\mathcal{U})$, then we denote by $[X]_A$ the equivalence class of X. The ***stalk*** of $\mathscr{F}$ over A is the set of all equivalence classes. The stalk of $\mathscr{F}$ over A is denoted by $\mathscr{F}_A$, and we write $\mathscr{F}_{\{x\}}$ as $\mathscr{F}_x$. In particular, $\mathscr{G}^{\nu}_{x,\mathsf{TM}}$ is the stalk at x of the sheaf $\mathscr{G}^{\nu}_{\mathsf{TM}}$ of C^{ν}-vector fields.

4.2 Sheafification

Let us now describe how a presheaf can be converted in a natural way into a sheaf. The description of how to do this for general presheaves is a little complicated. However, in the case we are dealing with here, we can be explicit about this.

Lemma 4.6 (A sheaf associated to every presheaf of sets of vector fields). *Let $m \in \mathbb{Z}_{\geq 0}$ and $m' \in \{0, \text{lip}\}$, let $\nu \in \{m+m', \infty, \omega\}$, and let $r \in \{\infty, \omega\}$, as required. Let M be a C^r-manifold and let $\mathscr{F}$ be a presheaf of sets of C^ν-vector fields. For an open set $\mathcal{U} \subseteq \mathsf{M}$, define*

$$\mathrm{Sh}(\mathscr{F})(\mathcal{U}) = \{X \in \Gamma^\nu(\mathsf{T}\mathcal{U}) \mid [X]_x \in \mathscr{F}_x \text{ for every } x \in \mathcal{U}\}.$$

Then $\mathrm{Sh}(\mathscr{F})$ is a sheaf.

Proof Let $\mathcal{U} \subseteq \mathsf{M}$ be open and let $(\mathcal{U}_a)_{a\in A}$ be an open cover of $\mathcal{U}$. Suppose that local sections $X_a \in \mathrm{Sh}(\mathscr{F})(\mathcal{U}_a)$, $a \in A$, satisfy $X_a|\mathcal{U}_a \cap \mathcal{U}_b = X_b|\mathcal{U}_a \cap \mathcal{U}_b$ for each $a, b \in A$. Since $\mathscr{G}^\nu_{\mathsf{TM}}$ is a sheaf, there exists $X \in \Gamma^\nu(\mathsf{T}\mathcal{U})$ such that $X|\mathcal{U}_a = X_a$, $a \in A$. It remains to show that $X \in \mathrm{Sh}(\mathscr{F})(\mathcal{U})$. Let $x \in \mathcal{U}$ and let $a \in A$ be such that $x \in \mathcal{U}_a$. Then we have $[X]_x = [X_a]_x \in \mathscr{F}_x$, as desired. □

With the lemma in mind we have the following definition.

Definition 4.7 (Sheafification of a presheaf of sets of vector fields). Let $m \in \mathbb{Z}_{\geq 0}$ and $m' \in \{0, \text{lip}\}$, let $\nu \in \{m+m', \infty, \omega\}$, and let $r \in \{\infty, \omega\}$, as required. Let M be a C^r-manifold and let $\mathscr{F}$ be a presheaf of sets of C^ν-vector fields. The ***sheafification*** of $\mathscr{F}$ is the sheaf $\mathrm{Sh}(\mathscr{F})$ of sets of vector fields defined by

$$\mathrm{Sh}(\mathscr{F})(\mathcal{U}) = \{X \in \Gamma^\nu(\mathsf{T}\mathcal{U}) \mid [X]_x \in \mathscr{F}_x \text{ for all } x \in \mathcal{U}\}.$$ ○

Let us consider some examples of sheafification.

Examples 4.8 (Sheafification).

1. Let us consider the presheaf of bounded vector fields from Example 4.3–1. Let $m \in \mathbb{Z}_{\geq 0}$ and $m' \in \{0, \text{lip}\}$, let $\nu \in \{m+m', \infty, \omega\}$, and let $r \in \{\infty, \omega\}$, as required. Let M be a C^r-manifold and consider the presheaf $\mathscr{F}_{\text{bdd}}$ of bounded vector fields. One easily sees that the stalk of this presheaf at $x \in \mathsf{M}$ is given by

$$\mathscr{F}_{\text{bdd},x} = \{[X]_x \mid X \in \Gamma^\nu(\mathsf{TM})\},$$

i.e., there are no restrictions on the stalks coming from the boundedness restriction on vector fields. Therefore, $\mathrm{Sh}(\mathscr{F}_{\text{bdd}}) = \mathscr{G}^\nu_{\mathsf{TM}}$.
2. Let us consider the sheafification of the presheaf defined in Example 4.3–2 by a family of vector fields defined on the open sets of an open cover of M. Thus we let $m \in \mathbb{Z}_{\geq 0}$ and $m' \in \{0, \text{lip}\}$, let $\nu \in \{m+m', \infty, \omega\}$, and let $r \in \{\infty, \omega\}$, as required. We let $(\mathcal{U}_a)_{a\in A}$ be an open cover for M and let $(X_a)_{a\in A}$ be C^ν-vector

fields defined on these sets. By $\mathscr{F}$ we denote the associated presheaf. Note that, for $x \in \mathsf{M}$,

$$\mathscr{F}_x = \{[X_a]_x \mid x \in \mathcal{U}_a\}.$$

Generally, the sheafification of this presheaf will be difficult to understand. However, in the case that $\mathcal{U}_a \cap \mathcal{U}_b = \emptyset$, then the vector field $X \in \Gamma^\nu(\mathsf{T}\mathcal{U})$, $\mathcal{U} = \mathcal{U}_a \cup \mathcal{U}_b$, defined by

$$X(x) = \begin{cases} X_a(x), & x \in \mathcal{U}_a, \\ X_b(x), & x \in \mathcal{U}_b \end{cases}$$

is a section of $\mathrm{Sh}(\mathscr{F})$ over $\mathcal{U}$.

3. Let us now examine the sheafification of a globally generated presheaf of sets of vector fields as in Example 4.3–3. Let $m \in \mathbb{Z}_{\geq 0}$ and $m' \in \{0, \mathrm{lip}\}$, let $\nu \in \{m + m', \infty, \omega\}$, and let $r \in \{\infty, \omega\}$, as required. Let M be a C^r-manifold and let $\mathscr{F}$ be a globally generated presheaf of sets of C^ν-vector fields, with $\mathscr{F}(\mathsf{M})$ the global generators. We will contrast $\mathscr{F}(\mathcal{U})$ with $\mathrm{Sh}(\mathscr{F})(\mathcal{U})$ to get an understanding of what the sheaf $\mathrm{Sh}(\mathscr{F})$ "looks like".
To do so, for $\mathcal{U} \subseteq \mathsf{M}$ open and for $X \in \Gamma^\nu(\mathsf{T}\mathcal{U})$, let us define a set-valued map $\kappa_{X,\mathcal{U}} \colon \mathcal{U} \twoheadrightarrow \mathscr{F}(\mathsf{M})$ by

$$\kappa_{X,\mathcal{U}}(x) = \{X' \in \mathscr{F}(\mathsf{M}) \mid X'(x) = X(x)\}.$$

Generally, since we have asked nothing of the vector field X, we might have $\kappa_{X,\mathcal{U}}(x) = \emptyset$ for a chosen x, or for some x, or for every x. If, however, we take $X \in \mathscr{F}(\mathcal{U})$, then $X = X'|\mathcal{U}$ for some $X' \in \mathscr{F}(\mathsf{M})$. Therefore, there exists a constant selection of $\kappa_{X,\mathcal{U}}$, i.e., a constant function $s \colon \mathcal{U} \to \mathscr{F}(\mathsf{M})$ such that $s(x) \in \kappa_{X,\mathcal{U}}(x)$ for every $x \in \mathcal{U}$. Note that if, for example, M is connected and $\nu = \omega$, then there will be a *unique* such constant selection since a real analytic vector field known on an open subset uniquely determines the vector field on the connected component containing this open set; this is the Identity Theorem, cf. [6, Theorem A.3] in the holomorphic case and the same proof applies in the real analytic case. Moreover, this constant selection in this case will completely characterise $\kappa_{X,\mathcal{U}}$ in the sense that $\kappa_{X,\mathcal{U}}(x) = \{s(x)\}$.
Let us now contrast this with the character of the map $\kappa_{X,\mathcal{U}}$ for a local section $X \in \mathrm{Sh}(\mathscr{F})(\mathcal{U})$. In this case, for each $x \in \mathcal{U}$, we have $[X]_x = [X_x]_x$ for some $X_x \in \mathscr{F}(\mathsf{M})$. Thus there exists a neighbourhood $\mathcal{V}_x \subseteq \mathcal{U}$ such that $X|\mathcal{V}_x = X_x|\mathcal{V}_x$. What this shows is that there is a locally constant selection of $\kappa_{X,\mathcal{U}}$, i.e., a locally constant map $s \colon \mathcal{U} \to \mathscr{F}(\mathsf{M})$ such that $s(x) \in \kappa_{X,\mathcal{U}}(x)$ for each $x \in \mathcal{U}$. As above, in the real analytic case when M is connected, this locally constant selection is uniquely determined, and determines $\kappa_{X,\mathcal{U}}$ in the sense that $\kappa_{X,\mathcal{U}}(x) = \{s(x)\}$. Note that locally constant functions are those that are constant on connected components. Thus, by passing to the sheafification, we have gained flexibility by allowing local sections to differ on connected components of an open set. While this does not completely characterise the difference between local sections of the

globally generated sheaf $\mathscr{F}$ and its sheafification $\mathrm{Sh}(\mathscr{F})$, it captures the essence of the matter, and *does* completely characterise the difference when $\nu = \omega$ and M is connected. ○

4.3 The Étalé Space

For the basic material on tautological control systems in Chap. 5, the elementary machinery provided above for presheaves and sheaves of sets of vector fields will suffice. However, as the theory of tautological control systems develops, it appears likely that more sophisticated sheaf theoretic constructions will play a rôle. We begin to see this in Chap. 6 where we introduce the machinery needed to define a class of trajectories extending those defined in Chap. 5. In the next two sections we shall introduce two sheaf theoretic constructions that we will need in Chap. 6.

The first has to do with a natural topology associated to a sheaf. We describe this in the following definitions, which are very restricted versions of general definitions that one uses in a complete presentation of sheaf theory.

Definition 4.9 (Étalé space, étalé topology). Let $m \in \mathbb{Z}_{\geq 0}$ and $m' \in \{0, \mathrm{lip}\}$, let $\nu \in \{m + m', \infty, \omega\}$, and let $r \in \{\infty, \omega\}$, as required. Let M be a C^r-manifold and let $\mathscr{F}$ be a presheaf of sets of C^ν-vector fields.

(i) The ***étalé space*** for the presheaf $\mathscr{F}$ is the disjoint union of the stalks:

$$\mathrm{Et}(\mathscr{F}) = \mathring{\bigcup_{x \in \mathsf{M}}} \mathscr{F}_x.$$

The ***étalé projection*** is the map $\pi_{\mathscr{F}} : \mathrm{Et}(\mathscr{F}) \to \mathsf{M}$ defined by $\pi_{\mathscr{F}}([X]_x) = x$.

(ii) The ***étalé topology*** for $\mathrm{Et}(\mathscr{F})$ is that topology for $\mathrm{Et}(\mathscr{F})$ generated by the basis

$$\mathcal{B}(\mathcal{U}, X) = \{[X]_x \mid x \in \mathcal{U}\}, \qquad \mathcal{U} \subseteq \mathsf{M} \text{ open}, \; X \in \mathscr{F}(\mathcal{U}).$$ ○

The étalé topology is an important construction in sheaf theory, and here we will review a few of the basic ideas and constructions associated with it.

Remarks 4.10 (Properties of the étalé space and the étalé topology). We let $m \in \mathbb{Z}_{\geq 0}$ and $m' \in \{0, \mathrm{lip}\}$, let $\nu \in \{m + m', \infty, \omega\}$, and let $r \in \{\infty, \omega\}$, as required, and we let $\mathscr{F}$ be a presheaf of sets of C^ν-vector fields on the C^r-manifold M.

1. The étalé projection is a local homeomorphism [12, Lemma 2.3.5(a)].
2. A ***local section*** of $\mathrm{Et}(\mathscr{F})$ over an open set $\mathcal{U}$ is a continuous map $\sigma : \mathcal{U} \to \mathrm{Et}(\mathscr{F})$ with the property that $\pi_{\mathscr{F}} \circ \sigma = \mathrm{id}_{\mathcal{U}}$.
3. If (and only if) $\mathscr{F}$ is a sheaf, there is a bijective correspondence between local sections of $\mathscr{F}$ over $\mathcal{U}$ and local sections of $\mathrm{Et}(\mathscr{F})$ defined by the map sending $X \in \mathscr{F}(\mathcal{U})$ to the local section σ_X of $\mathrm{Et}(\mathscr{F})$ given by $\sigma_X(x) = [X]_x$ [12, Lemma 2.4.3]. Thus we are really to think of these two flavours of local sections as being the same. Indeed, normally in sheaf theory there is no additional notation such as we use for $\mathrm{Et}(\mathscr{F})$, and $\mathscr{F}$ and $\mathrm{Et}(\mathscr{F})$ are typically identified.

4. The étalé topology for $\mathrm{Et}(\mathscr{G}^{\nu}_{\mathsf{TM}})$ is not Hausdorff for $\nu \in \{m + m', \infty\}$. Let us illustrate this for $\nu = \infty$. Let $\mathcal{U} \subseteq \mathsf{M}$ be open with a nonempty boundary and let $X \in \Gamma^{\infty}(\mathsf{TM})$ be a smooth vector field that is nowhere zero on $\mathcal{U}$ and is zero on $\mathsf{M}\setminus\mathcal{U}$. Let Z be the zero vector field. If $x \in \mathrm{bd}(\mathcal{U})$, we claim that any neighbourhoods of $[X]_x$ and $[Z]_x$ intersect. To see this, let $\mathcal{O}_X$ and $\mathcal{O}_Z$ be neighbourhoods in the étalé topology of $[X]_x$ and $[Z]_x$. Since any sufficiently small neighbourhood of $[X]_x$ and $[Z]_x$ is homeomorphic to a neighbourhood of x under the étalé projection, let us suppose without loss of generality that $\mathcal{O}_X$ and $\mathcal{O}_Z$ are both homeomorphic to a neighbourhood $\mathcal{V}$ of x under the projection. For $y \in \mathcal{V} \cap (\mathsf{M} \setminus \mathrm{cl}(\mathcal{U}))$, $[X]_y = [Z]_y$. Since $\mathcal{O}_X$ and $\mathcal{O}_Z$ are uniquely determined by the germs of X and Z in $\mathcal{V}$, respectively, it follows that $[X]_y = [Z]_y \in \mathcal{O}_X \cap \mathcal{O}_Z$, giving the desired conclusion.
5. The étalé topology for $\mathrm{Et}(\mathscr{G}^{\omega}_{\mathsf{TM}})$ is Hausdorff. Though this is well-known, we could not find a reference for this, so let us prove it. Let $[X]_x$ and $[Y]_y$ be distinct. If $x \neq y$ then there are disjoint neighbourhoods $\mathcal{U}$ and $\mathcal{V}$ of x and y and then $\mathcal{B}(\mathcal{U}, X)$ and $\mathcal{B}(\mathcal{V}, Y)$ are disjoint neighbourhoods of $[X]_x$ and $[Y]_y$, respectively, since the étalé projection is an homeomorphism from the neighbourhoods in M to the neighbourhoods in $\mathrm{Et}(\mathscr{G}^{\omega}_{\mathsf{TM}})$. If $x = y$ let $[X]_x$ and $[Y]_x$ be distinct and suppose that any neighbourhoods of $[X]_x$ and $[Y]_x$ in the étalé topology intersect. This implies, in particular, that, for every connected neighbourhood $\mathcal{U}$ of x, the basic neighbourhoods $\mathcal{B}(\mathcal{U}, X)$ and $\mathcal{B}(\mathcal{U}, Y)$ intersect. This implies the existence of an open subset $\mathcal{V}$ of $\mathcal{U}$ such that Y and Y agree on $\mathcal{V}$. This, however, contradicts the Identity Theorem, cf. [6, Theorem A.3]. Note that this implies that the étalé space of every presheaf of sets of real analytic vector fields is Hausdorff since $\mathrm{Et}(\mathscr{F}) \subseteq \mathrm{Et}(\mathscr{G}^{\omega}_{\mathsf{TM}})$. ○

4.4 Stalk Topologies

In order to characterise some of our constructions with tautological control systems in Chap. 5, we will make use of a convenient topology for the stalks of the sheaf of vector fields.

Definition 4.11 (C$^{\nu}$-stalk topology). Let $m \in \mathbb{Z}_{\geq 0}$ and $m' \in \{0, \mathrm{lip}\}$, let $\nu \in \{m + m', \infty, \omega\}$, let $r \in \{\infty, \omega\}$, as required, and let M be a C^r-manifold. For $x \in \mathsf{M}$, the **C$^{\nu}$*-stalk topology*** for $\mathscr{G}^{\nu}_{x,\mathsf{TM}}$ is the final topology[1] for the restriction mappings

$$\begin{aligned} r_{\mathcal{U},x}\colon \Gamma^{\nu}(\mathsf{T}\mathcal{U}) &\to \mathscr{G}^{\nu}_{x,\mathsf{TM}} \\ X &\mapsto [X]_x, \end{aligned}$$

where $\mathcal{U} \subseteq \mathsf{M}$ is a neighbourhood of x. ○

The following property of the stalk topology will be useful for us.

[1] That is to say, it is the finest locally convex topology for which the given mappings are continuous [2, §I.2.4].

Lemma 4.12 (The C^ν-stalk topology determines the C^ν-topology). *Let $m \in \mathbb{Z}_{\geq 0}$ and $m' \in \{0, \text{lip}\}$, let $\nu \in \{m + m', \infty, \omega\}$, and let $r \in \{\infty, \omega\}$, as required. For a manifold M of class C^r, the following topologies for $\Gamma^\nu(\mathsf{TM})$ agree:*

(i) *the C^ν-topology;*
(ii) *the initial topology[2] defined by the mappings $r_{\mathsf{M},x}$, $x \in \mathsf{M}$, and the C^ν-stalk topologies.*

Proof The definition of the C^ν-stalk topology ensures that the mappings $r_{\mathsf{M},x}, x \in \mathsf{M}$, are continuous if $\Gamma^\nu(\mathsf{TM})$ has the C^ν-topology. This means that the C^ν-topology is finer than the initial topology.

To show the converse, we show that every open set for $\Gamma^\nu(\mathsf{TM})$ in the C^ν-topology is open in the initial topology. We first show that, if $K \subseteq \mathsf{M}$ is compact, if p_K is one of the seminorms (2.2), if $\varepsilon \in \mathbb{R}_{>0}$, and if $X \in \Gamma^\nu(\mathsf{TM})$, then there is a neighbourhood of X in the initial topology contained in the subbasic open set

$$\mathcal{B}(X, K, \varepsilon) = \{Y \in \Gamma^\nu(\mathsf{TM}) \mid p_K(Y - X) < \varepsilon\}.$$

Thus let K, ε, and X be so chosen. Let $x \in K$ and let $(\mathcal{U}_{x,j})_{j\in\mathbb{Z}_{>0}}$ be a sequence of relatively compact neighbourhoods of x such that (1) $\text{cl}(\mathcal{U}_{x,j+1}) \subseteq \mathcal{U}_{x,j}$, $j \in \mathbb{Z}_{>0}$ and (2) $\cap_{j\in\mathbb{Z}_{>0}}\mathcal{U}_{x,j} = \{x\}$. Since the directed set (under inclusion) $(\mathcal{U}_{x,j})_{j\in\mathbb{Z}_{>0}}$ is cofinal in the directed set of all neighbourhoods of x, the topology of $\mathscr{G}^\nu_{x,\mathsf{TM}}$ is the direct limit topology of the sequence $(\Gamma^\nu(\mathsf{T}\mathcal{U}_{x,j})))_{j\in\mathbb{Z}_{>0}}$ under the mappings $r_{\mathcal{U}_{x,j},x}$ [5, page 137]. A neighbourhood base of 0 in $\mathscr{G}^\nu_{x,\mathsf{TM}}$ is thus given by

$$\text{conv}\Big(\bigcup_{j\in\mathbb{Z}_{>0}} r_{\mathcal{U}_{x,j}}(\mathcal{O}_{x,j})\Big), \tag{4.1}$$

where $\mathcal{O}_{x,j}$ are neighbourhoods of $0 \in \Gamma^\nu(\mathsf{T}\mathcal{U}_{x,j})$, $j \in \mathbb{Z}_{>0}$, [5, Proposition 4.1.1]. Let us specifically take

$$\mathcal{O}_{x,j} = \{Z \in \Gamma^\nu(\mathsf{T}\mathcal{U}_{x,j}) \mid p_{\text{cl}(\mathcal{U}_{x,j+1})}(Z) < \varepsilon\}$$

and define $\mathcal{O}_x$ to be the corresponding convex hull, as in (4.1). Now let $x_1, \dots, x_k \in K$ be such that $K \subseteq \cup_{l=1}^k \mathcal{U}_{x_l,1}$. We claim that

$$\bigcap_{l=1}^k r_{\mathcal{U},x_l}^{-1}(X + \mathcal{O}_{x_l}) \subseteq \mathcal{B}(X, K, \varepsilon).$$

If $Y \in \cap_{l=1}^k r_{\mathcal{U},x_l}^{-1}(X + \mathcal{O}_{x_l})$ then, for each $l \in \{1, \dots, k\}$,

[2] That it to say, it is the coarsest topology for which the given mappings are continuous [2, §I.2.3].

$$r_{\mathcal{U},x_l}(Y-X) \in \mathcal{O}_{x_l} \implies Y - X = \sum_{a_l=1}^{r_l} \lambda_{l,a_l}[Z_{l,a_l}]_{x_l},$$

where $\lambda_{l,a_1}, \dots, \lambda_{l,a_{r_l}} \in [0,1]$ satisfy $\sum_{a_l=1}^{r_l} \lambda_{l,a_l} = 1$ and where $Z_{l,a_l} \in \Gamma^\nu(\mathsf{T}\mathcal{U}_{x_l,j_{a_l}})$ satisfies $p_{\mathrm{cl}(\mathcal{U}_{x_l,j_{a_l}+1})}(Z_{l,a_l}) < \varepsilon$ for some $j_{a_1}, \dots, j_{a_{r_l}} \in \mathbb{Z}_{>0}$. Therefore,

$$p_{\mathrm{cl}(\mathcal{U}_{x_l,j_{a_l}+1})}\Big(\sum_{a_l=1}^{r_l} \lambda_{l,a_l} Z_{l,a_l}\Big) \le \sum_{a_l=1}^{r_l} \lambda_{l,a_l} p_{\mathrm{cl}(\mathcal{U}_{x_l,j_{a_l}+1})}(Z_{l,a_l}) < \varepsilon.$$

Since $K \subseteq \cup_{l=1}^{k} \cup_{a_l=1}^{r_l} \mathrm{cl}(\mathcal{U}_{x_l,j_{a_l}+1})$, it follows that $p_K(Y-X) < \varepsilon$, as claimed. Now, since each of the sets $r_{\mathsf{M},x_l}^{-1}(X + \mathcal{O}_{x_l})$, $l \in \{1, \dots, k\}$, is open in the initial topology of $\Gamma^\nu(\mathsf{TM})$, we conclude that their intersection is a neighbourhood of X in the initial topology contained in $\mathcal{B}(X, K, \varepsilon)$.

To complete the proof, if $\mathcal{O} \subseteq \Gamma^\nu(\mathsf{TM})$ is a neighbourhood of $X \in \Gamma^\nu(\mathsf{TM})$, then there are $K_1, \dots, K_s \subseteq \mathsf{M}$ compact and $\varepsilon_1, \dots, \varepsilon_s \in \mathbb{R}_{>0}$ (and whatever other pieces of data may be required to define the s basic seminorms of the form (2.2)) such that

$$\bigcap_{j=1}^{s} \mathcal{B}(X, K_j, \varepsilon_j) \subseteq \mathcal{O}.$$

By the arguments above, we have neighbourhoods $\mathcal{O}_1, \dots, \mathcal{O}_s$ of X in the initial topology such that $\mathcal{O}_j \subseteq \mathcal{B}(X, K_j, \varepsilon_j)$, $j \in \{1, \dots, s\}$. Then $\cap_{j=1}^{s} \mathcal{O}_j \subseteq \mathcal{O}$ is a neighbourhood of X in the initial topology. □

The C^ν-stalk topology has a rather different character in the case of $\nu = \omega$ than the other cases. The following result enumerates some of these differences.

Lemma 4.13 (Properties of the stalk topology). *Let $m \in \mathbb{Z}_{\ge 0}$ and $m' \in \{0, \mathrm{lip}\}$, let $\nu \in \{m + m', \infty, \omega\}$, and let $r \in \{\infty, \omega\}$, as required. Let M be a C^r-manifold having no zero-dimensional connected components. Then the following two statements hold for each $x \in \mathsf{M}$:*

(i) the closure of $\{0\} \subseteq \mathscr{G}^\nu_{x,\mathsf{TM}}$ is

$$\begin{cases} \{[X]_x \in \mathscr{G}^m_{x,\mathsf{TM}} \mid j_m X(x) = 0_x\}, & \nu = m, \\ \{[X]_x \in \mathscr{G}^{m+\mathrm{lip}}_{x,\mathsf{TM}} \mid j_m X(x) = 0_x,\ \mathrm{dil}\, j_m X(x) = 0\}, & \nu = m + \mathrm{lip}, \\ \{[X]_x \in \mathscr{G}^\infty_{x,\mathsf{TM}} \mid j_m X(x) = 0_x,\ m \in \mathbb{Z}_{\ge 0}\}, & \nu = \infty, \\ \{0\}, & \nu = \omega; \end{cases}$$

(ii) the C^ν-stalk topology for $\mathscr{G}^\nu_{x,\mathsf{TM}}$ is Hausdorff if and only if $\nu = \omega$;

(iii) for $\mathcal{U} \subseteq \mathsf{M}$ open and connected and for $x \in \mathcal{U}$, the mapping $r_{\mathcal{U},x}$ is injective if and only if $\nu = \omega$.

Proof (i) We consider the various cases.

$\nu = m$: First of all, if $j_m X(x) \neq 0$, then $[X]_x$ is not in the closure of $\{0\}$. Indeed, if $j_m X(x) \neq 0$ then $\|j_m X(x)\|_{\overline{\mathbb{G}}_m} > 0$. Therefore, if $\mathcal{U}$ is any neighbourhood of x on which X is defined and if $K \subseteq \mathcal{U}$ is compact with $x \in K$, $p^m_K(X) > 0$. Thus, since $\Gamma^m(\mathsf{T}\mathcal{U})$ is Hausdorff, there exists a neighbourhood $\mathcal{O}$ of 0 in $\Gamma^m(\mathsf{T}\mathcal{U})$ such that $X \notin \mathcal{O}$. It follows that $[X]_x$ is not in any neighbourhood of 0 in $\mathscr{G}^m_{x,\mathsf{TM}}$ containing $r_{\mathcal{U},x}(\mathcal{O})$. That is, $[X]_x \notin \mathrm{cl}(\{0\})$.

Let $(\mathcal{U}, \phi)$ be a chart about x such that $\overline{\mathsf{B}}(1, \phi(x)) \subseteq \phi(\mathcal{U})$. For simplicity, suppose that $\phi(x) = \mathbf{0}$. Let $X \in \Gamma^m(\mathsf{T}\mathcal{U})$ be such that $j_m X(x) = 0_x$. Let $f \in \mathrm{C}^\infty(\mathbb{R}^n)$ be such that $f(\mathbf{x}) = 0$ for $\mathbf{x}$ in a neighbourhood of $\mathbf{0}$ and such that $f(\mathbf{x}) = 1$ for $\mathbf{x} \notin \overline{\mathsf{B}}(1,\mathbf{0})$. Let

$$X_j(y) = f(j\phi(y))X(y), \qquad j \in \mathbb{Z}_{\geq 0},\ y \in \mathcal{U}.$$

Note that $[X_j]_x = 0$. We claim that $(X_j)_{j\in\mathbb{Z}_{>0}}$ converges to X in the C^m-topology of $\Gamma^m(\mathsf{T}\mathcal{U})$. We will show, in fact, that $(j_m X_j)_{j\in\mathbb{Z}_{>0}}$ converges uniformly to $j_m X$ on $\mathcal{U}$. Indeed, let $\varepsilon \in \mathbb{R}_{>0}$, let $\mathbf{X}$ be the local representative of X and let $\mathbf{X}_j$ be the local representative of X_j. For $s \in \{0, 1, \dots, m\}$ and $\mathbf{x} \in \phi(\mathcal{U})$, we compute

$$\|\boldsymbol{D}^s(\mathbf{X} - \mathbf{X}_j)(\mathbf{x})\| \leq \sum_{r=0}^{s} \frac{s!}{r!(s-r)!} j^r \|\boldsymbol{D}^{s-r}\mathbf{X}(\mathbf{x})\| \|\boldsymbol{D}^r(1-f)(j\mathbf{x})\|.$$

Note that $\boldsymbol{D}^r(1-f)(j\mathbf{x}) = 0$ for $\mathbf{x} \in \phi(\mathcal{U}) \setminus \overline{\mathsf{B}}(\frac{1}{j}, \mathbf{0})$ and $r \in \{0, 1, \dots, m\}$. By the binomial theorem,

$$\sum_{r=0}^{s} \frac{s!}{r!(s-r)!} = 2^s.$$

Therefore, for $\varepsilon \in \mathbb{R}_{>0}$, if we choose $N \in \mathbb{Z}_{>0}$ sufficiently large that

$$\boldsymbol{D}^s\mathbf{X}(\mathbf{x}) \leq \frac{\varepsilon}{(2k)^m \|\boldsymbol{D}^r(1-f)\|_\infty}, \qquad r, s \in \{0, 1, \dots, m\},$$

for all $\mathbf{x} \in \overline{\mathsf{B}}(\frac{1}{N}, \mathbf{0})$, then

$$\|\boldsymbol{D}^s(\mathbf{X} - \mathbf{X}_j)(\mathbf{x})\| < \varepsilon, \qquad \mathbf{x} \in \phi(\mathcal{U}),\ s \in \{0, 1, \dots, m\},\ j \geq N.$$

giving the desired uniform convergence.

Let $\mathcal{U}$ and X be as in the preceding paragraph. Note that $r^{-1}_{\mathcal{U},x}(\{0\})$ is a submodule of $\Gamma^m(\mathsf{T}\mathcal{U})$; it is the submodule of vector fields vanishing in some neighbourhood of x. Our argument from the preceding paragraph shows that $X \in \mathrm{cl}(r^{-1}_{\mathcal{U},x}(\{0\}))$. Therefore, $[X]_x \in \mathrm{cl}(\{0\})$ by [13, Theorem 7.2(d)].

$\nu = m + \mathrm{lip}$: We consider the case $m = 0$ since the general case follows along similar lines, with more complicated notation.

First of all, if $X(x) \neq 0$ then, as in the previous part of the proof, we can conclude that $[X]_x \notin \mathrm{cl}\left(\mathscr{G}^{\mathrm{lip}}_{x,\mathsf{TM}}\right)$. Next suppose that $X(x) = 0$ but that $\mathrm{dil}\, X(x) \neq 0$. This means that, for any neighbourhood $\mathcal{U}$ of x on which X is defined and any compact $K \subseteq \mathcal{U}$ with $x \in K$, $\lambda^0_K(X) > 0$. Since $\Gamma^{\mathrm{lip}}(\mathsf{T}\mathcal{U})$ is Hausdorff, there exists a neighbourhood $\mathcal{O}$ of $0 \in \Gamma^{\mathrm{lip}}(\mathsf{T}\mathcal{U})$ such that $X \notin \mathcal{O}$. From this we conclude that $[X]_x$ is not in any neighbourhood of $0 \in \mathscr{G}^{\mathrm{lip}}_{x,\mathsf{TM}}$ containing $r_{\mathcal{U},x}(\mathcal{O})$. Thus $[X]_x \notin \mathrm{cl}(\{0\})$.

Now let $X \in \Gamma^{\mathrm{lip}}(\mathsf{TM})$ be such that $X(x) = 0$ and $\mathrm{dil}\, X(x) = 0$. Let $(\mathcal{U}, \phi)$ be a chart about x with $\phi(x) = \mathbf{0}$ and with $\overline{\mathsf{B}}(1, \mathbf{0}) \subseteq \phi(\mathcal{U})$. Define a sequence $(X_j)_{j\in\mathbb{Z}_{>0}}$ as in the preceding part of the proof. As in that part of the proof, the sequence $(X_j)_{j\in\mathbb{Z}_{>0}}$ converges uniformly to X on $\mathcal{U}$. We also compute

$$\|\mathbf{X}(\mathbf{x}) - \mathbf{X}_j(\mathbf{x}) - (\mathbf{X}(\mathbf{y}) - \mathbf{X}_j(\mathbf{y}))\| = \|(f(j\mathbf{x}) - 1)\mathbf{X}(\mathbf{x}) - (f(j\mathbf{y}) - 1)\mathbf{X}(\mathbf{y})\|.$$

Let $\varepsilon \in \mathbb{R}_{>0}$ and let $N \in \mathbb{Z}_{>0}$ be sufficiently large that

$$\|\mathbf{X}(\mathbf{x})\| \leq \frac{\varepsilon}{4\|1 - f\|_\infty}$$

for $\mathbf{x} \in \overline{\mathsf{B}}(\frac{1}{N}, \mathbf{0})$. Let $j \geq N$. We have four cases.

1. $\mathbf{x}, \mathbf{y} \in \overline{\mathsf{B}}(\frac{1}{N}, \mathbf{0})$: In this case we have
$$\|\mathbf{X}(\mathbf{x}) - \mathbf{X}_j(\mathbf{x}) - (\mathbf{X}(\mathbf{y}) - \mathbf{X}_j(\mathbf{y}))\| \leq 2\|1 - f\|_\infty(\|\mathbf{X}(\mathbf{x})\| + \|\mathbf{X}(\mathbf{y})\|) \leq \varepsilon.$$
2. $\mathbf{x} \in \overline{\mathsf{B}}(\frac{1}{N}, \mathbf{0})$, $\mathbf{y} \in \phi(\mathcal{U}) \setminus \mathsf{B}(\frac{1}{N}, \mathbf{0})$: Here we have $1 - f(j\mathbf{x}) = 0$, and so we estimate
$$\|\mathbf{X}(\mathbf{x}) - \mathbf{X}_j(\mathbf{x})\| \leq \|1 - f\|_\infty \|\mathbf{X}(\mathbf{x})\| < \varepsilon.$$
3. $\mathbf{x} \in \phi(\mathcal{U}) \setminus \mathsf{B}(\frac{1}{N}, \mathbf{0})$, $\mathbf{y} \in \overline{\mathsf{B}}(\frac{1}{N}, \mathbf{0})$: This is the same as the previous case.
4. $\mathbf{x}, \mathbf{y} \in \phi(\mathcal{U}) \setminus \mathsf{B}(\frac{1}{N}, \mathbf{0})$: In this case, we have
$$\|\mathbf{X}(\mathbf{x}) - \mathbf{X}_j(\mathbf{x}) - (\mathbf{X}(\mathbf{y}) - \mathbf{X}_j(\mathbf{y}))\| = 0.$$

This shows that $(X_j)_{j\in\mathbb{Z}_{>0}}$ converges to X in the $\mathrm{C}^{\mathrm{lip}}$-topology.

As in the final paragraph of the preceding part of the proof, with $\mathcal{U}$ and X as in the preceding paragraph, $X \in \mathrm{cl}(r^{-1}_{\mathcal{U},x}(\{0\}))$ and so $[X]_x \in \mathrm{cl}(\{0\})$.

$\nu = \infty$: If $j_m X(x) \neq 0$ for some $m \in \mathbb{Z}_{\geq 0}$, then we may argue as in the case $\nu = m$ to see that $[X]_x$ is not in $\mathrm{cl}(\{0\})$. If $j_m X(x) = 0$ for every $m \in \mathbb{Z}_{\geq 0}$, then the computations above in the case $\nu = m$, applied to each $m \in \mathbb{Z}_{\geq 0}$, give a sequence $(X_j)_{j\in\mathbb{Z}_{>0}}$ converging to X in the C^∞-topology on a neighbourhood $\mathcal{U}$ of x. The same argument as in the case $\nu = m$ then gives $[X]_x \in \mathrm{cl}(\{0\})$.

$\nu = \omega$: Let $[X]_x \in \mathscr{G}^{\omega}_{x,\mathsf{TM}}$ be nonzero. This is equivalent to the infinite jet of X at x being nonzero. One can infer from this, using the arguments in the smooth case above, that $[X]_x$ is not in the closure of $\{0\}$ in the C^∞-stalk topology. Now we note that the C^ω-topology is finer that the C^∞-topology, and so the same holds for the stalk topologies. Thus closed sets for the C^ω-stalk topology are also closed sets for the C^∞-stalk topology, and we conclude that $[X]_x$ is not in the closure of $\{0\}$ in the C^ω-stalk topology.

(ii) This follows immediately from part (i).
(iii) If $\nu = \omega$, then the Identity Theorem [6, Theorem A.3] implies that, if $r_{\mathcal{U},x}(X) = 0$, then $X = 0$ since X must vanish in some neighbourhood of x. If $\nu \neq \omega$, then we let $\mathcal{V}, \mathcal{V}' \subseteq \mathcal{U}$ be a relatively compact neighbourhoods of x such that

$$\mathrm{cl}(\mathcal{V}) \subseteq \mathcal{V}' \subseteq \mathrm{cl}(\mathcal{V}') \subseteq \mathcal{U}.$$

Then let $f \in \mathrm{C}^\infty(\mathsf{M})$ be such that f has the value 0 in some neighbourhood of $\mathrm{cl}(\mathcal{V})$ and has value 1 outside $\mathrm{cl}(\mathcal{V}')$ [1, Proposition 5.5.8]. Now, for any $X, Y \in \Gamma^\nu(\mathsf{T}\mathcal{U})$,

$$r_{\mathcal{U},x}(X) = r_{\mathcal{U},x}(X + fY).$$

This prohibits injectivity of $r_{\mathcal{U},x}$. □

References

1. Abraham R, Marsden JE, Ratiu TS (1988) Manifolds, tensor analysis, and applications, 2nd edn. No. 75 in Applied Mathematical Sciences. Springer, Berlin
2. Bourbaki N (1989) General topology I. Elements of Mathematics. Springer, Berlin
3. Bredon GE (1997) Sheaf theory, 2nd edn. No. 170 in Graduate Texts in Mathematics. Springer, New York
4. Godement R (1958) Topologie algébrique et théorie des faisceaux. No. 13 in Publications de l'Institut de mathématique de l'Université de Strasbourg. Hermann, Paris
5. Groethendieck A (1973) Topological vector spaces. Notes on Mathematics and its Applications. Gordon & Breach Science Publishers, New York
6. Gunning RC (1990) Introduction to holomorphic functions of several variables, vol I: function theory. Wadsworth & Brooks/Cole Mathematical Series. Wadsworth & Brooks/Cole, Belmont, CA
7. Kashiwara M, Schapira P (1990) Sheaves on manifolds. No. 292 in Grundlehren der Mathematischen Wissenschaften. Springer, Berlin
8. Ramanan S (2005) Global calculus. No. 65 in Graduate Studies in Mathematics. American Mathematical Society, Providence, RI
9. Stacks Project Authors (2014) Stacks Project. http://stacks.math.columbia.edu
10. Stefan P (1974) Accessible sets, orbits and foliations with singularities. Proc London Math Soc 29:699–713
11. Sussmann HJ (1973) Orbits of families of vector fields and integrability of distributions. Trans Am Math Soc 180:171–188
12. Tennison BR (1976) Sheaf theory. No. 20 in London Mathematical Society Lecture Note Series. Cambridge University Press, New York
13. Willard S (2004) General topology. Dover Publications Inc., New York. Reprint of 1970 Addison-Wesley edition

Chapter 5
Tautological Control Systems: Definitions and Fundamental Properties

In this chapter we introduce the class of control systems, tautological control systems, that we propose as being useful mathematical models for the investigation of geometric system structure. As promised in our introduction in Sect. 1.2, this class of systems naturally handles a variety of regularity classes; we work with finitely differentiable, Lipschitz, smooth, and real analytic classes simultaneously with comparative ease. Also as indicated in Sect. 1.2, the framework makes essential use of sheaf theory in its formulation. We shall see in Sect. 5.6 that the natural morphisms for tautological control systems ensure feedback-invariance of the theory.

5.1 Tautological Control Systems

Our definition of a tautological control system is relatively straightforward, given the constructions of the preceding chapter.

Definition 5.1 (Tautological control system and related notions). Let $m \in \mathbb{Z}_{\geq 0}$ and $m' \in \{0, \text{lip}\}$, let $\nu \in \{m + m', \infty, \omega\}$, and let $r \in \{\infty, \omega\}$, as required.

(i) A **C^ν*-tautological control system*** is a pair $\mathfrak{G} = (\mathsf{M}, \mathscr{F})$, where M is a manifold of class C^r whose elements are called ***states*** and where $\mathscr{F}$ is a presheaf of sets of C^ν-vector fields on M.

(ii) A tautological control system $\mathfrak{G} = (\mathsf{M}, \mathscr{F})$ is ***complete*** if $\mathscr{F}$ is a sheaf and is ***globally generated*** if $\mathscr{F}$ is globally generated.

(iii) The ***completion*** of $\mathfrak{G} = (\mathsf{M}, \mathscr{F})$ is the tautological control system $\mathrm{Sh}(\mathfrak{G}) = (\mathsf{M}, \mathrm{Sh}(\mathscr{F}))$. ○

This is a pretty featureless definition, sorely in need of some connection to control theory. Let us begin to build this connection by pointing out the manner in which more common constructions give rise to tautological control systems, and vice versa.

A.D. Lewis, *Tautological Control Systems*, SpringerBriefs in Control, Automation and Robotics, DOI: 10.1007/978-3-319-08638-5_5

Examples 5.2 (Correspondences between tautological control systems and other sorts of control systems). One of the topics of interest to us will be the relationship between our notion of tautological control systems and the more common notions of control systems (as in Sects. 3.3.1 and 3.3.2) and differential inclusions (as in Sect. 3.3.3). We begin here by making some more or less obvious associations.

1. Let $m \in \mathbb{Z}_{\geq 0}$ and $m' \in \{0, \text{lip}\}$, let $\nu \in \{m + m', \infty, \omega\}$, and let $r \in \{\infty, \omega\}$, as required. Let $\Sigma = (\mathsf{M}, F, \mathcal{C})$ be a C^ν-control system. To this control system we associate the C^ν-tautological control system $\mathfrak{G}_\Sigma = (\mathsf{M}, \mathscr{F}_\Sigma)$ by

$$\mathscr{F}_\Sigma(\mathcal{U}) = \{F^u|\mathcal{U} \in \Gamma^\nu(\mathsf{T}\mathcal{U}) \mid u \in \mathcal{C}\}.$$

   The presheaf of sets of vector fields in this case is of the globally generated variety, as in Example 4.3–3. According to Example 4.3–3, we should generally not expect tautological control systems such as this to be a priori complete. We can, however, sheafify so that the tautological control system $\mathrm{Sh}(\mathfrak{G}_\Sigma)$ is complete.
2. Let us consider a means of going from a large class of tautological control systems to a control system. Let $m \in \mathbb{Z}_{\geq 0}$ and $m' \in \{0, \text{lip}\}$, let $\nu \in \{m + m', \infty, \omega\}$, and let $r \in \{\infty, \omega\}$, as required. We suppose that we have a C^ν-tautological control system $\mathfrak{G} = (\mathsf{M}, \mathscr{F})$ where the presheaf $\mathscr{F}$ is globally generated. We define a C^ν-control system $\Sigma_\mathfrak{G} = (\mathsf{M}, F_\mathscr{F}, \mathcal{C}_\mathscr{F})$ as follows. We take $\mathcal{C}_\mathscr{F} = \mathscr{F}(\mathsf{M})$, i.e., the control set is our family of globally defined vector fields and the topology is that induced from $\Gamma^\nu(\mathsf{TM})$. We define

$$\begin{aligned} F_\mathscr{F} \colon \mathsf{M} \times \mathcal{C}_\mathscr{F} &\to \mathsf{TM} \\ (x, X) &\mapsto X(x). \end{aligned}$$

   (Note that one has to make an awkward choice between writing a vector field as u or a control as X, since vector fields are controls. We have gone with the latter awkward choice, since it more readily mandates thinking about what the symbols mean.) Note that $F_\mathscr{F}^X = X$, and so this is somehow the identity map in disguise. In order for this construction to provide a bona fide control system, we should check that $F_\mathscr{F}$ is a parameterised vector field of class C^ν according to Theorem 3.18. For this it is sufficient to check that the map $X \mapsto F_\mathscr{F}^X$ is continuous. But this is the identity map, which is obviously continuous!
   Note that $\Sigma_\mathfrak{G}$ is a control-linear system, according to Example 3.23.
3. Let $m \in \mathbb{Z}_{\geq 0}$ and $m' \in \{0, \text{lip}\}$, let $\nu \in \{m + m', \infty, \omega\}$, and let $r \in \{\infty, \omega\}$, as required. Let $\mathscr{X} : \mathsf{M} \twoheadrightarrow \mathsf{TM}$ be a differential inclusion. If $\mathcal{U} \subseteq \mathsf{M}$ is open, we denote

$$\Gamma^\nu(\mathscr{X}|\mathcal{U}) = \{X \in \Gamma^\nu(\mathsf{T}\mathcal{U}) \mid X(x) \in \mathscr{X}(x),\ x \in \mathcal{U}\}.$$

   One should understand, of course, that we may very well have $\Gamma^\nu(\mathscr{X}|\mathcal{U}) = \emptyset$. This might happen for two reasons.

(a) First, the differential inclusion may lack sufficient regularity to permit even local sections of the prescribed regularity.
(b) Second, even if it permits local sections, there may be be problems finding sections defined on "large" open sets, because there may be global obstructions. One might anticipate this to be especially problematic in the real analytic case, where the specification of a vector field locally determines its behaviour globally by the Identity Theorem, cf. [11, Theorem A.3].

This caveat notwithstanding, we can go ahead and define a tautological control system $\mathfrak{G}_{\mathscr{X}} = (\mathsf{M}, \mathscr{F}_{\mathscr{X}})$ with $\mathscr{F}_{\mathscr{X}}(\mathcal{U}) = \Gamma^{\nu}(\mathscr{X}|\mathcal{U})$.
We claim that $\mathfrak{G}_{\mathscr{X}}$ is complete. To see this, let $\mathcal{U} \subseteq \mathsf{M}$ be open and let $(\mathcal{U}_a)_{a \in A}$ be an open cover for $\mathcal{U}$. For each $a \in A$, let $X_a \in \mathscr{F}_{\mathscr{X}}(\mathcal{U}_a)$ and suppose that, for $a, b \in A$,

$$X_a|\mathcal{U}_a \cap \mathcal{U}_b = X_b|\mathcal{U}_a \cap \mathcal{U}_b.$$

Since $\mathscr{G}^{\nu}_{\mathsf{TM}}$ is a sheaf, let $X \in \Gamma^{\nu}(\mathsf{T}\mathcal{U})$ be such that $X|\mathcal{U}_a = X_a$ for each $a \in A$. We claim that $X \in \mathscr{F}_{\mathscr{X}}(\mathcal{U})$. Indeed, for $x \in \mathcal{U}$ we have $X(x) = X_a(x) \in \mathscr{X}(x)$ if we take $a \in A$ such that $x \in \mathcal{U}_a$.
The sheaf $\mathscr{F}_{\mathscr{X}}$ is not often globally generated since it is, indeed, a sheaf as we saw in Example 4.3–3. Here is a stupid counterexample that relates to the character of differential inclusions. Let us define $\mathscr{X}(x) = \mathsf{T}_x\mathsf{M}$, $x \in \mathsf{M}$, so that $\mathscr{F}_{\mathscr{X}} = \mathscr{G}^{\nu}_{\mathsf{TM}}$. For an open set $\mathcal{U}$, there will generally be local sections $X \in \Gamma^{\nu}(\mathsf{T}\mathcal{U})$ that are not restrictions to $\mathcal{U}$ of globally defined vector fields; vector fields that "blow up" at some point in the boundary of $\mathcal{U}$ are what one should have in mind.
4. Let $m \in \mathbb{Z}_{\geq 0}$ and $m' \in \{0, \text{lip}\}$, let $\nu \in \{m + m', \infty, \omega\}$, and let $r \in \{\infty, \omega\}$, as required. Note that there is also associated to any C^{ν}-tautological control system $\mathfrak{G} = (\mathsf{M}, \mathscr{F})$ a differential inclusion $\mathscr{X}_{\mathfrak{G}}$ by

$$\mathscr{X}_{\mathfrak{G}}(x) = \{X(x) \mid [X]_x \in \mathscr{F}_x\},$$

recalling that $\mathscr{F}_x$ is the stalk of $\mathscr{F}$ at x. ∘

Now note that we can iterate the four constructions and ask to what extent we end up back where we started. More precisely, we have the following result.

Proposition 5.3 (Going back and forth between classes of systems). *Let $m \in \mathbb{Z}_{\geq 0}$ and $m' \in \{0, \text{lip}\}$, let $\nu \in \{m + m', \infty, \omega\}$, and let $r \in \{\infty, \omega\}$, as required. Let $\mathfrak{G} = (\mathsf{M}, \mathscr{F})$ be a C^{ν}-tautological control system, let $\Sigma = (\mathsf{M}, F, \mathcal{C})$ be a C^{ν}-control system, and let $\mathscr{X}$ be a differential inclusion. Then the following statements hold:*

(i) if $\mathfrak{G}$ is globally generated, then $\mathfrak{G}_{\Sigma_{\mathfrak{G}}} = \mathfrak{G}$;
(ii) if the map $u \mapsto F^u$ from $\mathcal{C}$ to $\Gamma^{\nu}(\mathsf{TM})$ is injective and open onto its image, then $\Sigma_{\mathfrak{G}_{\Sigma}} = \Sigma$;
(iii) $\mathscr{F}(\mathcal{U}) \subseteq \mathscr{F}_{\mathscr{X}_{\mathfrak{G}}}(\mathcal{U})$ for every open $\mathcal{U} \subseteq \mathsf{M}$;
(iv) $\mathscr{X}_{\mathfrak{G}_{\mathscr{X}}} \subseteq \mathscr{X}$.

Proof (i) Let $\mathcal{U} \subseteq \mathsf{M}$ be open and let $X \in \mathscr{F}(\mathcal{U})$. Then $X = X'|\mathcal{U}$ for $X' \in \mathscr{F}(\mathsf{M})$. Thus $X' \in \mathcal{C}_{\mathscr{F}}$ and $X'(x) = F(x, X')$ and so $X \in \mathscr{F}_{\Sigma_{\mathfrak{G}}}(\mathcal{U})$. Conversely, let $X \in \mathscr{F}_{\Sigma_{\mathfrak{G}}}(\mathcal{U})$. Then $X(x) = F(x, X')$, $x \in \mathcal{U}$, for some $X' \in \mathcal{C}_{\mathscr{F}}$. But this means that $X(x) = X'(x)$ for $X' \in \mathscr{F}(\mathcal{U})$ and for all $x \in \mathcal{U}$. In other words, $X \in \mathscr{F}(\mathcal{U})$.

(ii) Note that $\mathfrak{G}_\Sigma$ is globally generated. Thus we have

$$\mathcal{C}_{\mathscr{F}_\Sigma} = \mathscr{F}_\Sigma(\mathsf{M}) = \{F^u \mid u \in \mathcal{C}\}.$$

Since the map $u \mapsto F^u$ is continuous, and injective and open onto its image (by hypothesis), it is an homeomorphism onto its image. Thus $\mathcal{C}_{\mathscr{F}_\Sigma}$ is homeomorphic to $\mathcal{C}$. Since $u \mapsto F^u$ is injective we can unambiguously write

$$F_{\mathscr{F}_\Sigma}(x, F^u) = F^u(x) = F(x, u).$$

(iii) Let $\mathcal{U} \subseteq \mathsf{M}$ be open. If $X \in \mathscr{F}(\mathcal{U})$, then clearly we have $X(x) \in \mathscr{X}_{\mathfrak{G}}(x)$ for every $x \in \mathcal{U}$ and so $\mathscr{F}(\mathcal{U}) \subseteq \mathscr{F}_{\mathscr{X}_{\mathfrak{G}}}(\mathcal{U})$, giving the assertion.

(iv) This is obvious. □

Remark 5.4 (Correspondence between control systems and control-linear systems). The result establishes the rather surprising correspondence between control systems $\Sigma = (\mathsf{M}, F, \mathcal{C})$ for which the map $u \mapsto F^u$ is injective and open onto its image, and the associated control-linear system $\Sigma_{\mathfrak{G}_\Sigma} = (\mathsf{M}, \mathscr{F}_{\Sigma_{\mathfrak{G}}}, \mathcal{C}_{\mathscr{F}_\Sigma})$. That is to say, at least at the system level, in our treatment every system corresponds in a natural way to a control-linear system, albeit with a rather complicated control set. This correspondence carries over to trajectories as well, but one can also weaken these conditions to obtain trajectory correspondence in more general situations. These matters we discuss in detail in Sect. 5.5. ○

Let us make some comments on the hypotheses present in the preceding result.

Remarks 5.5 (Going back and forth between classes of systems).

1. Since $\mathfrak{G}_\Sigma$ is necessarily globally generated for any control system Σ, the requirement that $\mathfrak{G}$ be globally generated cannot be dropped in part (i).
2. The requirement that the map $u \mapsto F^u$ be injective in part (ii) cannot be relaxed. Without this assumption, there is no way to recover F from $\{F^u \mid u \in \mathcal{C}\}$. Similarly, if this map is not open onto its image, while there may be a bijection between $\mathcal{C}$ and $\mathcal{C}_{\mathscr{F}_\Sigma}$, it will not be an homeomorphism which one needs for the control systems to be the same.
3. The converse assertion in part (iii) does not generally hold, as many counterexamples show. Here are two, each of a different character.

   (a) We take $\mathsf{M} = \mathbb{R}$ and consider the C^ω-tautological control system $\mathfrak{G} = (\mathsf{M}, \mathscr{F})$ where $\mathscr{F}$ is the globally generated presheaf defined by the single vector field $x^2 \frac{\partial}{\partial x}$. Note that

$$\mathscr{X}_{\mathfrak{G}}(x) = \begin{cases} \{0\}, & x = 0, \\ \mathsf{T}_x\mathbb{R}, & x \neq 0. \end{cases}$$

Therefore,

$$\mathscr{F}_{\mathscr{X}}(\mathcal{U}) = \begin{cases} \{X \in \Gamma^\omega(\mathsf{T}\mathcal{U}) \mid X(0) = 0\}, & 0 \in \mathcal{U}, \\ \Gamma^\omega(\mathsf{T}\mathcal{U}), & 0 \notin \mathcal{U}. \end{cases}$$

It holds, therefore, that the vector field $x\frac{\partial}{\partial x}$ is a global section of $\mathscr{F}_{\mathscr{X}}$, but is not a global section of $\mathscr{F}$.

(b) Let us again take $\mathsf{M} = \mathbb{R}$ and now define a smooth tautological control system $\mathfrak{G} = (\mathsf{M}, \mathscr{F})$ by asking that $\mathscr{F}$ be the globally generated presheaf defined by the vector fields $X_1, X_2 \in \Gamma^\infty(\mathbb{R})$, where

$$X_1(x) = \begin{cases} \mathrm{e}^{-1/x}\frac{\partial}{\partial x}, & x > 0, \\ 0, & x \leq 0, \end{cases}$$

and

$$X_2(x) = \begin{cases} \mathrm{e}^{-1/x}\frac{\partial}{\partial x}, & x < 0, \\ 0, & x \geq 0. \end{cases}$$

In this case,

$$\mathscr{X}_{\mathfrak{G}}(x) = \begin{cases} \{0\}, & x = 0, \\ \{0\} \cup \{\mathrm{e}^{-1/x}\frac{\partial}{\partial x}\}, & x \neq 0. \end{cases}$$

Therefore, $\mathscr{F}_{\mathscr{X}}$ is the sheafification of the globally generated presheaf defined by the vector fields X_1, X_2, X_3, and X_4, where

$$X_3(x) = \begin{cases} \mathrm{e}^{-1/x}\frac{\partial}{\partial x}, & x \neq 0, \\ 0, & x = 0, \end{cases}$$

and X_4 is the zero vector field.

4. Given the discussion in Example 5.2–3, one cannot reasonably expect that we will generally have equality in part (iv) of the preceding result. Indeed, one might even be inclined to say that it is only differential inclusions satisfying $\mathscr{X} = \mathscr{X}_{\mathfrak{G}_{\mathscr{X}}}$ that are useful in geometric control theory… ○

While we are not yet finished with the task of formulating our theory—trajectories have yet to appear—it is worthwhile to make a pause at this point to reflect upon what we have done and have not done. After a moments thought, one realises that the difference between a control system $\Sigma = (\mathsf{M}, F, \mathcal{C})$ and its associated tautological control system $\mathfrak{G}_\Sigma = (\mathsf{M}, \mathscr{F}_\Sigma)$ is that, in the former case, the control vector fields are from the *indexed family* $(F^u)_{u \in \mathcal{C}}$, while for the tautological control system we have the *set* $\{F^u \mid u \in \mathcal{C}\}$. In going from the former to the latter we have "forgotten" the index u which we are explicitly keeping track of for control systems. If the map $u \mapsto F^u$ is injective, as in Proposition 5.3(ii), then there is no information lost as one goes from the indexed family to the set. If $u \mapsto F^u$ is not injective, then this is a signal that the control set is too large, and perhaps one should collapse it in some way. In other words, one can probably suppose injectivity of $u \mapsto F^u$ without loss of generality. (Openness of this map is another matter. As we shall see in Sect. 5.5 below, openness (and a little more) is crucial for there to be trajectory correspondence between systems and tautological control systems.) This then leaves us with the mathematical semantics of distinguishing between the indexed family $(F^u)_{u \in \mathcal{C}}$ and the subset $\{F^u \mid u \in \mathcal{C}\}$. About this, let us make two observations.

1. The entire edifice of nonlinear control theory seems, in some sense, to be built upon the preference of the indexed family over the set. As we discuss in Chap. 1, in applications there are very good reasons for doing this. But from the point of view of the general theory, the idea that one should carefully maintain the labelling of the vector fields from the set $\{F^u \mid u \in \mathcal{C}\}$ seems to be a really unnecessary distraction. And, moreover, it is a distraction upon which is built the whole notion of "feedback transformation", plus entire methodologies in control theory that are not feedback-invariant, e.g., linearisation, cf. Example 1.1. So, semantics? Possibly, but sometimes semantic choices are important.
2. Many readers will probably not be convinced by our attempts to magnify the distinction between the indexed family $(F^u)_{u \in \mathcal{C}}$ and the set $\{F^u \mid u \in \mathcal{C}\}$. As we shall see, however, this distinction becomes more apparent if one is really dedicated to using sets rather than indexed families. Indeed, this deprives one of the notion of "control", and one is forced to be more thoughtful about what one means by "trajectory".

We comment that the preceding discussion only pertains to the comparison of globally generated tautological control systems and "ordinary" control systems. Although we have not yet seen the significance of this, this sort of discussion essentially ignores the "presheaf of vector fields" aspect of tautological control systems, and this is likely to feature heavily in further developments of tautological control systems, cf. Chaps. 6 and 7.

5.2 Open-Loop Systems

Trajectories are associated to "open-loop systems", so we first discuss these. We first introduce some notation. Let $m \in \mathbb{Z}_{\geq 0}$ and $m' \in \{0, \mathrm{lip}\}$, let $\nu \in \{m + m', \infty, \omega\}$, and let $r \in \{\infty, \omega\}$, as required. For a C^ν-tautological control system $\mathfrak{G} = (\mathsf{M}, \mathscr{F})$, we then denote

$$\mathrm{LI}\Gamma^\nu(\mathbb{T}; \mathscr{F}(\mathcal{U})) = \{X : \mathbb{T} \to \mathscr{F}(\mathcal{U}) \mid X \in \mathrm{LI}\Gamma^\nu(\mathbb{T}; (\mathsf{T}\mathcal{U}))\},$$

for $\mathbb{T} \subseteq \mathbb{R}$ an interval and $\mathcal{U} \subseteq \mathsf{M}$ open.

Definition 5.6 (Open-loop system). Let $m \in \mathbb{Z}_{\geq 0}$ and $m' \in \{0, \mathrm{lip}\}$, let $\nu \in \{m + m', \infty, \omega\}$, and let $r \in \{\infty, \omega\}$, as required. Let $\mathfrak{G} = (\mathsf{M}, \mathscr{F})$ be a C^ν-tautological control system. An ***open-loop system*** for $\mathfrak{G}$ is a triple $\mathfrak{G}_{\mathrm{ol}} = (X, \mathbb{T}, \mathcal{U})$ where

(i) $\mathbb{T} \subseteq \mathbb{R}$ is an interval called the ***time-domain***;
(ii) $\mathcal{U} \subseteq \mathsf{M}$ is open;
(iii) $X \in \mathrm{LI}\Gamma^\nu(\mathbb{T}; \mathscr{F}\mathcal{U})$. ○

Note that an open-loop system for $\mathfrak{G} = (\mathsf{M}, \mathscr{F})$ is also an open-loop system for the completion $\mathrm{Sh}(\mathfrak{G})$, just because $\mathscr{F}(\mathcal{U}) \subseteq \mathrm{Sh}(\mathscr{F})(\mathcal{U})$. However, of course, there may be open-loop systems for $\mathrm{Sh}(\mathfrak{G})$ that are not open-loop systems for $\mathfrak{G}$. This is as it should be, and has no significant ramifications for the theory, at least for the purposes of this chapter.

In order to see how we should think about an open-loop system, let us consider this notion in the special case of control systems.

Example 5.7 (Open-loop systems associated to control systems). Let $m \in \mathbb{Z}_{\geq 0}$ and $m' \in \{0, \mathrm{lip}\}$, let $\nu \in \{m + m', \infty, \omega\}$, and let $r \in \{\infty, \omega\}$, as required. Let $\Sigma = (\mathsf{M}, F, \mathcal{C})$ be a C^ν-control system with $\mathfrak{G}_\Sigma$ the associated C^ν-tautological control system. If we let $\mu \in \mathrm{L}^{\mathrm{cpt}}_{\mathrm{loc}}(\mathbb{T}; \mathcal{C})$, then we have the associated open-loop system $\mathfrak{G}_{\Sigma,\mu} = (F^\mu, \mathbb{T}, \mathsf{M})$ defined by

$$F^\mu(t)(x) = F(x, \mu(t)), \qquad t \in \mathbb{T},\ x \in \mathsf{M}.$$

Proposition 3.20 ensures that this is an open-loop system for the tautological control system $\mathfrak{G}_\Sigma$.

A similar assertion holds if $\mathcal{C}$ is a subset of a locally convex topological vector space and F defines a sublinear control system, and if $\mu \in \mathrm{L}^1_{\mathrm{loc}}(\mathbb{T}; \mathcal{C})$, cf. Proposition 3.24. ○

Notation 5.8 (Open-loop systems). For an open-loop system $\mathfrak{G}_{\mathrm{ol}}(X, \mathbb{T}, \mathcal{U})$, the notation $X(t)(x)$, while accurate, is unnecessarily cumbersome, and we will often instead write $X(t, x)$ or $X_t(x)$, with no loss of clarity and a gain in aesthetics. ○

Generally one might wish to place a restriction on the set of open-loop systems one will use. This is tantamount to, for usual control systems, placing restrictions on the controls one might use; one may wish to use piecewise continuous controls or piecewise constant controls, for example. For tautological control systems we do this as follows.

Definition 5.9 (Open-loop subfamily). Let $m \in \mathbb{Z}_{\geq 0}$ and $m' \in \{0, \mathrm{lip}\}$, let $\nu \in \{m + m', \infty, \omega\}$, and let $r \in \{\infty, \omega\}$, as required. Let $\mathfrak{G} = (\mathsf{M}, \mathscr{F})$ be a C^ν-tautological control system. An ***open-loop subfamily*** for $\mathfrak{G}$ is an assignment, to each interval $\mathbb{T} \subseteq \mathbb{R}$ and each open set $\mathcal{U} \subseteq \mathsf{M}$, a subset $\mathscr{O}_{\mathfrak{G}}(\mathbb{T}, \mathcal{U}) \subseteq \mathrm{LI}\Gamma^\nu(\mathbb{T}; \mathscr{F}(\mathcal{U}))$ with the property that, if $(\mathbb{T}_1, \mathcal{U}_1)$ and $(\mathbb{T}_2, \mathcal{U}_2)$ are such that $\mathbb{T}_1 \subseteq \mathbb{T}_2$ and $\mathcal{U}_1 \subseteq \mathcal{U}_2$, then

$$\{X|\mathbb{T}_1 \times \mathcal{U}_1 \mid X \in \mathscr{O}_{\mathfrak{G}}(\mathbb{T}_2, \mathcal{U}_2)\} \subseteq \mathscr{O}_{\mathfrak{G}}(\mathbb{T}_1, \mathcal{U}_1).$$ ○

Here are a few common examples of open-loop subfamilies.

Examples 5.10 (Open-loop subfamilies). Let $m \in \mathbb{Z}_{\geq 0}$ and $m' \in \{0, \mathrm{lip}\}$, let $\nu \in \{m + m', \infty, \omega\}$, and let $r \in \{\infty, \omega\}$, as required. Let $\mathfrak{G} = (\mathsf{M}, \mathscr{F})$ be a C^ν-tautological control system.

1. The ***full subfamily*** for $\mathfrak{G}$ is the open-loop subfamily $\mathscr{O}_{\mathfrak{G},\mathrm{full}}$ defined by

$$\mathscr{O}_{\mathfrak{G},\mathrm{full}}(\mathbb{T}, \mathcal{U}) = \mathrm{LI}\Gamma^\nu(\mathbb{T}; \mathscr{F}(\mathcal{U})).$$

Thus the full subfamily contains all possible open-loop systems. Of course, every open-loop subfamily will be contained in this one.

2. The ***locally essentially bounded subfamily*** for $\mathfrak{G}$ is the open-loop subfamily $\mathscr{O}_{\mathfrak{G},\infty}$ defined by asking that

$$\mathscr{O}_{\mathfrak{G},\infty}(\mathbb{T}, \mathcal{U}) = \{X \in \mathscr{O}_{\mathfrak{G},\mathrm{full}}(\mathbb{T}, \mathcal{U}) \mid X \in \mathrm{LB}\Gamma^\nu(\mathbb{T}; \mathsf{T}\mathcal{U})\}.$$

Thus, for the locally essentially bounded subfamily, we require that the condition of being locally integrally C^ν-bounded be replaced with the stronger condition of being locally essentially C^ν-bounded.

3. The ***locally essentially compact subfamily*** for $\mathfrak{G}$ is the open-loop subfamily $\mathscr{O}_{\mathfrak{G},\mathrm{cpt}}$ defined by asking that

$$\mathscr{O}_{\mathfrak{G},\mathrm{cpt}}(\mathbb{T}, \mathcal{U}) = \{X \in \mathscr{O}_{\mathfrak{G},\mathrm{full}}(\mathbb{T}, \mathcal{U}) \mid \text{ for every compact subinterval } \mathbb{T}' \subseteq \mathbb{T} \text{ there exists a compact } K \subseteq \Gamma^\nu(\mathbb{T}; \mathsf{T}\mathcal{U}) \text{ such that } X(t) \subseteq K \text{ for almost every } t \in \mathbb{T}'\}.$$

Thus, for the locally essentially compact subfamily, we require that the condition of being locally essentially bounded in the von Neumann bornology (that defines the locally essentially bounded subfamily) be replaced with being locally essentially bounded in the compact bornology.

We comment that in cases when the compact and von Neumann bornologies agree, then of course we have $\mathscr{O}_{\mathfrak{G},\infty} = \mathscr{O}_{\mathfrak{G},\mathrm{cpt}}$. As pointed out by Jafarpour and Lewis [15], this is the case when $\nu \in \{\infty, \omega\}$.

4. The ***piecewise constant subfamily*** for $\mathfrak{G}$ is the open-loop subfamily $\mathscr{O}_{\mathfrak{G},\mathrm{pwc}}$ defined by asking that

$$\mathscr{O}_{\mathfrak{G},\mathrm{pwc}}(\mathbb{T}; \mathcal{U}) = \{X \in \mathscr{O}_{\mathfrak{G},\mathrm{full}}(\mathbb{T}, \mathcal{U}) \mid t \mapsto X(t) \text{ is piecewise constant}\}.$$

Let us be clear what we mean by piecewise constant. We mean that there is a partition $(\mathbb{T}_j)_{j \in J}$ of $\mathbb{T}$ into pairwise disjoint intervals such that

   (a) for any compact interval $\mathbb{T}' \subseteq \mathbb{T}$, the set

$$\{j \in J \mid \mathbb{T}' \cap \mathbb{T}_j \neq \emptyset\}$$

is finite and such that
(b) $X|\mathbb{T}_j$ is constant for each $j \in J$.

One might imagine that the piecewise constant open-loop subfamily will be useful for studying orbits and controllability of tautological control systems. This is almost true, but one needs the more refined notion of an étalé trajectory considered in Chap. 6.

5. We can associate an open-loop subfamily to an open-loop system as follows. Let $\mathscr{O}_{\mathfrak{G}}$ be an open-loop subfamily for $\mathfrak{G}$, let $\mathbb{T}$ be a time-domain, let $\mathcal{U} \subseteq \mathsf{M}$ be open, and let $X \in \mathscr{O}_{\mathfrak{G}}(\mathbb{T}, \mathcal{U})$. We denote by $\mathscr{O}_{\mathfrak{G},X}$ the open-loop subfamily defined as follows. If $\mathbb{T}' \subseteq \mathbb{T}$ and $\mathcal{U}' \subseteq \mathcal{U}$, then we let

$$\mathscr{O}_{\mathfrak{G},X}(\mathbb{T}', \mathcal{U}') = \{X' \in \mathscr{O}_{\mathfrak{G}}(\mathbb{T}', \mathcal{U}') \mid X' = X|\mathbb{T}' \times \mathcal{U}'\}.$$

If $\mathbb{T}' \not\subseteq \mathbb{T}$ and/or $\mathcal{U}' \not\subseteq \mathcal{U}$, then we take $\mathscr{O}_{\mathfrak{G},X}(\mathbb{T}', \mathcal{U}') = \emptyset$. Thus $\mathscr{O}_{\mathfrak{G},X}$ is comprised of those vector fields from $\mathscr{O}_{\mathfrak{G}}$ that are merely restrictions of X to smaller domains.

6. In Proposition 5.3 we saw that there was a pretty robust correspondence between C^ν-control systems and C^ν-tautological control systems, *at the system level*. As we make our way towards trajectories, as we are now doing, this robustness breaks down a little. To frame this, we can define an open-loop subfamily for the tautological control system associated to a C^ν-control system $\Sigma = (\mathsf{M}, F, \mathcal{C})$ as follows. For a time-domain $\mathbb{T}$ and an open $\mathcal{U} \subseteq \mathsf{M}$, we define

$$\mathscr{O}_\Sigma(\mathbb{T}, \mathcal{U}) = \{F^\mu|\mathcal{U} \mid \mu \in \mathrm{L}^{\mathrm{cpt}}_{\mathrm{loc}}(\mathbb{T}; \mathcal{C})\},$$

recalling that $F^\mu(t, x) = F(x, \mu(t))$. We clearly have $\mathscr{O}_\Sigma(\mathbb{T}; \mathcal{U}) \subseteq \mathscr{O}_{\mathfrak{G}_\Sigma,\mathrm{cpt}}(\mathbb{T}; \mathcal{U})$ for every time-domain $\mathbb{T}$ and every open $\mathcal{U} \subseteq \mathsf{M}$; this is proved in the course of proving Proposition 3.20. (Of course, by virtue of Proposition 3.24, we have a corresponding construction if the control set $\mathcal{C}$ is a subset of a locally convex topological vector space, if F is sublinear, and if $\mu \in \mathrm{L}^1_{\mathrm{loc}}(\mathbb{T}; \mathcal{C})$.) However, we do not generally expect to have equality of the two open-loop subfamilies $\mathscr{O}_\Sigma$ and $\mathscr{O}_{\mathfrak{G}_\Sigma,\mathrm{cpt}}$. This, in turn, will have repercussions on the nature of the trajectories for these subfamilies, and, therefore, on the relationship of trajectories of a control system with those of the corresponding tautological control system. We will consider these matters in Sect. 5.5, and we will see that, for many interesting classes of control systems, there is, in fact, a natural trajectory correspondence between the system and its associated tautological control system. ∘

Our notion of an open-loop subfamily is very general, and working with the full generality will typically lead to annoying problems. There are many attributes that one may wish for open-loop subfamilies to satisfy in order to relax some the annoyance. To illustrate, let us define a typical attribute that one may require, that of translation-invariance. Let us define some notation so that we can easily make the definition. For a time-domain $\mathbb{T}$ and for $s \in \mathbb{R}$, we denote

$$s + \mathbb{T} = \{s + t \mid t \in \mathbb{T}\}$$

and we denote by $\tau_s : s + \mathbb{T} \to \mathbb{T}$ the translation map $\tau_s(t) = t - s$.

Definition 5.11 (Translation-invariant open-loop subfamily). Let $m \in \mathbb{Z}_{\geq 0}$ and $m' \in \{0, \mathrm{lip}\}$, let $\nu \in \{m + m', \infty, \omega\}$, and let $r \in \{\infty, \omega\}$, as required. Let $\mathfrak{G} = (\mathsf{M}, \mathscr{F})$ be a C^ν-tautological control system. An open-loop subfamily $\mathscr{O}_{\mathfrak{G}}$ for $\mathfrak{G}$ is ***translation-invariant*** if, for every $s \in \mathbb{R}$, every time-domain $\mathbb{T}$, and every open set $\mathcal{U} \subseteq \mathsf{M}$, the map

$$\begin{aligned}(\tau_s \times \mathrm{id}_{\mathcal{U}})^* : \mathscr{O}_{\mathfrak{G}}(s + \mathbb{T}, \mathcal{U}) &\to \mathscr{O}_{\mathfrak{G}}(\mathbb{T}, \mathcal{U}) \\ X &\mapsto X \circ (\tau_s \times \mathrm{id}_{\mathcal{U}})\end{aligned}$$

is a bijection. ○

An immediate consequence of the definition is, of course, that if $t \mapsto \xi(t)$ is a trajectory (we will formally define the notion of "trajectory" in the next section), then so is $t \mapsto \xi(s + t)$ for every $s \in \mathbb{R}$.

Let us now think about how open-loop subfamilies interact with completion. In order for the definition we are about to make make sense, we should verify the following lemma.

Lemma 5.12 (Time-varying vector fields characterised by their germs). *Let $m \in \mathbb{Z}_{\geq 0}$ and $m' \in \{0, \mathrm{lip}\}$, let $\nu \in \{m + m', \infty, \omega\}$, and let $r \in \{\infty, \omega\}$, as required. Let M be a C^r-manifold, let $\mathbb{T} \subseteq \mathbb{R}$ be an interval, and let $X : \mathbb{T} \times \mathsf{M} \to \mathsf{TM}$ have the property that $X(t, x) \in \mathsf{T}_x\mathsf{M}$ for each $(t, x) \in \mathbb{T} \times \mathsf{M}$. Then the following statements hold:*

(i) if, for each $x \in \mathsf{M}$, there exist a neighbourhood $\mathcal{U}$ of x and $X' \in \mathrm{CF}\Gamma^\nu(\mathbb{T}; \mathsf{T}\mathcal{U})$ such that $[X_t]_x = [X'_t]_x$ for every $t \in \mathbb{T}$, then $X \in \mathrm{CF}\Gamma^\nu(\mathbb{T}; \mathsf{TM})$;

(ii) if, for each $x \in \mathsf{M}$, there exist a neighbourhood $\mathcal{U}$ of x and $X' \in \mathrm{LI}\Gamma^\nu(\mathbb{T}; \mathsf{T}\mathcal{U})$ such that $[X_t]_x = [X'_t]_x$ for every $t \in \mathbb{T}$, then $X \in \mathrm{LI}\Gamma^\nu(\mathbb{T}; \mathsf{TM})$;

(iii) if, for each $x \in \mathsf{M}$, there exist a neighbourhood $\mathcal{U}$ of x and $X' \in \mathrm{LB}\Gamma^\nu(\mathbb{T}; \mathsf{T}\mathcal{U})$ such that $[X_t]_x = [X'_t]_x$ for every $t \in \mathbb{T}$, then $X \in \mathrm{LB}\Gamma^\nu(\mathbb{T}; \mathsf{TM})$.

Proof (i) Let $x \in \mathsf{M}$. Since X agrees in some neighbourhood of x with a Carathéodory vector field X', it follows that $t \mapsto X_t(x) = X'_t(x)$ is measurable. In like manner, let $t \in \mathbb{T}$ and let $x_0 \in \mathsf{M}$. Then $x \mapsto X_t(x) = X'_t(x)$ is of class C^ν in a neighbourhood of x_0, and so $x \mapsto X_t(x)$ is of class C^ν.

(ii) For $K \subseteq \mathsf{M}$ compact, for $k \in \mathbb{Z}_{\geq 0}$, and for $\mathbf{a} \in c_0(\mathbb{Z}_{\geq 0}; \mathbb{R}_{>0})$, denote

$$p_K = \begin{cases} p^\infty_{K,k}, & \nu = \infty, \\ p^m_K, & \nu = m, \\ p^{m+\mathrm{lip}}_K, & \nu = m + \mathrm{lip}, \\ p^\omega_{K,\mathbf{a}}, & \nu = \omega. \end{cases}$$

Let $K \subseteq \mathsf{M}$ be compact, let $x \in K$, let $\mathcal{U}_x$ be a relatively compact neighbourhood of x, and let $X_x \in \mathrm{LI}\Gamma^\nu(\mathbb{T}; \mathcal{U}_x)$ be such that $[X_t]_x = [X_{x,t}]_x$ for every $t \in \mathbb{T}$. By Lemma 3.10 there exists $g_x \in \mathrm{L}^1_{\mathrm{loc}}(\mathbb{T}; \mathbb{R}_{\geq 0})$ such that

$$p_{\mathrm{cl}(\mathcal{U}_x)}(X_{x,t}) \leq g_x(t), \quad t \in \mathbb{T}.$$

Now let $x_1, \dots, x_k \in K$ be such that $K \subseteq \cup_{j=1}^k \mathcal{U}_{x_j}$. Let $g(t) = \max\{g_{x_1}(t), \dots, g_{x_k}(t)\}$, noting that the associated function g is measurable by [7, Proposition 2.1.3] and is locally integrable by the triangle inequality, along with the fact that

$$g(t) \leq (g_{x_1}(t) + \cdots + g_{x_k}(t)).$$

We then have

$$p_K(X_t) \leq g(t), \quad t \in \mathbb{T},$$

showing that $X \in \mathrm{LI}\Gamma^\nu(\mathbb{T}; \mathsf{TM})$ by another application of Lemma 3.10.

(iii) This is proved in exactly the same manner, mutatis mutandis, as the preceding part of the lemma. □

The following definition can now be made.

Definition 5.13 (Completion of an open-loop subfamily). Let $m \in \mathbb{Z}_{\geq 0}$ and $m' \in \{0, \mathrm{lip}\}$, let $\nu \in \{m + m', \infty, \omega\}$, and let $r \in \{\infty, \omega\}$, as required. Let $\mathfrak{G} = (\mathsf{M}, \mathscr{F})$ be a C^ν-tautological control system and let $\mathscr{O}_{\mathfrak{G}}$ be an open-loop subfamily for $\mathfrak{G}$. The ***completion*** of $\mathscr{O}_{\mathfrak{G}}$ is the open-loop subfamily $\mathrm{Sh}(\mathscr{O}_{\mathfrak{G}})$ for $\mathrm{Sh}(\mathfrak{G})$ defined by specifying that $(X, \mathbb{T}, \mathcal{U}) \in \mathrm{Sh}(\mathscr{O}_{\mathfrak{G}})$ if, for each $x \in \mathcal{U}$, there exist a neighbourhood $\mathcal{U}' \subseteq \mathcal{U}$ of x and $(X', \mathbb{T}, \mathcal{U}') \in \mathscr{O}_{\mathfrak{G}}(\mathbb{T}, \mathcal{U}')$ such that $[X_t]_x = [X'_t]_x$ for each $t \in \mathbb{T}$. ∘

Clearly the completion of an open-loop subfamily is an open-loop subfamily for the completion. Moreover, if $(X, \mathbb{T}, \mathcal{U}) \in \mathscr{O}_{\mathfrak{G}}(\mathbb{T}, \mathcal{U})$, then $(X, \mathbb{T}, \mathcal{U}) \in \mathrm{Sh}(\mathscr{O}_{\mathfrak{G}}(\mathbb{T}, \mathcal{U}))$, but one cannot expect the converse assertion to generally hold.

5.3 Trajectories

With the concept of open-loop system just developed, it is relatively easy to provide a notion of a trajectory for a tautological control system.

Definition 5.14 (Trajectory for tautological control system). Let $m \in \mathbb{Z}_{\geq 0}$ and $m' \in \{0, \mathrm{lip}\}$, let $\nu \in \{m + m', \infty, \omega\}$, and let $r \in \{\infty, \omega\}$, as required. Let $\mathfrak{G} = (\mathsf{M}, \mathscr{F})$ be a C^ν-tautological control system and let $\mathscr{O}_{\mathfrak{G}}$ be an open-loop subfamily for $\mathfrak{G}$.

(i) For a time-domain $\mathbb{T}$, an open set $\mathcal{U} \subseteq \mathsf{M}$, and for $X \in \mathscr{O}_{\mathfrak{G}}(\mathbb{T}, \mathcal{U})$, an $(X, \mathbb{T}, \mathcal{U})$-***trajectory*** for $\mathscr{O}_{\mathfrak{G}}$ is a curve $\xi : \mathbb{T} \to \mathcal{U}$ such that $\xi'(t) = X(t, \xi(t))$ for almost every $t \in \mathbb{T}$.

(ii) For a time-domain $\mathbb{T}$ and an open set $\mathcal{U} \subseteq \mathsf{M}$, a $(\mathbb{T}, \mathcal{U})$-***trajectory*** for $\mathscr{O}_{\mathfrak{G}}$ is a curve $\xi : \mathbb{T} \to \mathcal{U}$ such that $\xi'(t) = X(t, \xi(t))$ for almost every $t \in \mathbb{T}$ for some $X \in \mathscr{O}_{\mathfrak{G}}(\mathbb{T}, \mathcal{U})$.

(iii) A ***trajectory*** for $\mathscr{O}_{\mathfrak{G}}$ is a curve that is a $(\mathbb{T}, \mathcal{U})$-trajectory for $\mathscr{O}_{\mathfrak{G}}$ for some time-domain $\mathbb{T}$ and some open set $\mathcal{U} \subseteq \mathsf{M}$.

We denote by:

(iv) $\mathrm{Traj}(X, \mathbb{T}, \mathcal{U}, \mathscr{O}_{\mathfrak{G}})$ the set of $(X, \mathbb{T}, \mathcal{U})$-trajectories for $\mathscr{O}_{\mathfrak{G}}$;

(v) $\mathrm{Traj}(\mathbb{T}, \mathcal{U}, \mathscr{O}_{\mathfrak{G}})$ the set of $(\mathbb{T}, \mathcal{U})$-trajectories for $\mathscr{O}_{\mathfrak{G}}$;

(vi) $\mathrm{Traj}(\mathscr{O}_{\mathfrak{G}})$ the set of trajectories for $\mathscr{O}_{\mathfrak{G}}$.

We shall abbreviate $\mathrm{Traj}(\mathbb{T},\mathcal{U},\mathfrak{G}) = \mathrm{Traj}(\mathbb{T},\mathcal{U},\mathscr{O}_{\mathfrak{G},\mathrm{full}})$ and $\mathrm{Traj}(\mathfrak{G}) = \mathrm{Traj}(\mathscr{O}_{\mathfrak{G},\mathrm{full}})$.

∘

Sometimes one wishes to keep track of the fact that, associated with a trajectory, is an open-loop system. The following notion is designed to capture this.

Definition 5.15 (Referenced trajectory). Let $m \in \mathbb{Z}_{\geq 0}$ and $m' \in \{0, \mathrm{lip}\}$, let $\nu \in \{m + m', \infty, \omega\}$, and let $r \in \{\infty, \omega\}$, as required. Let $\mathfrak{G} = (\mathsf{M}, \mathscr{F})$ be a C^{ν}-tautological control system and let $\mathscr{O}_{\mathfrak{G}}$ be an open-loop subfamily for $\mathfrak{G}$. A ***referenced*** $\mathscr{O}_{\mathfrak{G}}$***-trajectory*** is a pair (X, ξ) where $X \in \mathscr{O}_{\mathfrak{G}}(\mathbb{T}; \mathcal{U})$ and $\xi \in \mathrm{Traj}(X, \mathbb{T}, \mathcal{U})$. We denote by:

(i) $\mathrm{Rtraj}(\mathbb{T}, \mathcal{U}, \mathscr{O}_{\mathfrak{G}})$ the set of referenced $\mathscr{O}_{\mathfrak{G}}$-trajectories for which $X \in \mathscr{O}_{\mathfrak{G}}(\mathbb{T}; \mathcal{U})$;

(ii) $\mathrm{Rtraj}(\mathscr{O}_{\mathfrak{G}})$ the set of referenced trajectories. ∘

In Sect. 5.5 below, we shall explore trajectory correspondences between tautological control systems, control systems, and differential inclusions.

The notion of a trajectory immediately gives rise to a certain open-loop subfamily.

Example 5.16 (The open-loop subfamily defined by a trajectory). Let $m \in \mathbb{Z}_{\geq 0}$ and $m' \in \{0, \mathrm{lip}\}$, let $\nu \in \{m + m', \infty, \omega\}$, and let $r \in \{\infty, \omega\}$, as required. Let $\mathfrak{G} = (\mathsf{M}, \mathscr{F})$ be a C^{ν}-tautological control system, let $\mathscr{O}_{\mathfrak{G}}$ be an open-loop subfamily for $\mathfrak{G}$, and let $\xi \in \mathrm{Traj}(\mathbb{T}, \mathcal{U}, \mathscr{O}_{\mathfrak{G}})$. We denote by $\mathscr{O}_{\mathfrak{G},\xi}$ the open-loop subfamily defined as follows. If $\mathbb{T}' \subseteq \mathbb{T}$ and $\mathcal{U}' \subseteq \mathcal{U}$ are such that $\xi(\mathbb{T}') \subseteq \mathcal{U}'$, then we let

$$\mathscr{O}_{\mathfrak{G},\xi}(\mathbb{T}', \mathcal{U}') = \{X \in \mathscr{O}_{\mathfrak{G}}(\mathbb{T}', \mathcal{U}') \mid \xi'(t) = X(t, \xi(t)),\ \text{a.e. } t \in \mathbb{T}'\}.$$

If $\mathbb{T}' \not\subseteq \mathbb{T}$ or $\mathcal{U}' \not\subseteq \mathcal{U}$, or if $\mathbb{T}' \subseteq \mathbb{T}$ and $\mathcal{U}' \subseteq \mathcal{U}$ but $\xi(\mathbb{T}') \not\subseteq \mathcal{U}'$, then we take $\mathscr{O}_{\mathfrak{G},\xi}(\mathbb{T}', \mathcal{U}') = \emptyset$. Thus $\mathscr{O}_{\mathfrak{G},\xi}$ is comprised of those vector fields from $\mathscr{O}_{\mathfrak{G}}$ possessing ξ (restricted to the appropriate subinterval) as an integral curve. ∘

In control theory, trajectories are of paramount importance, often far more important, say, than systems per se. For this reason, one might ask that completion of a tautological control system preserve trajectories. However, this will generally not be the case, as the following counterexample illustrates.

Example 5.17 (Sheafification does not preserve trajectories). We will chat our way through a general example; the reader can very easily create a specific concrete instance from the general discussion.

Let $m \in \mathbb{Z}_{\geq 0}$ and $m' \in \{0, \text{lip}\}$, let $\nu \in \{m + m', \infty, \omega\}$, and let $r \in \{\infty, \omega\}$, as required. We let M be a C^r-manifold with Riemannian metric $\mathbb{G}$. We consider the presheaf $\mathscr{F}_{\mathsf{bdd}}$ of bounded C^ν-vector fields on M, initially discussed in Example 4.3–1. We let $\mathfrak{G}_{\mathsf{bdd}} = (\mathsf{M}, \mathscr{F}_{\mathsf{bdd}})$ so that, as we saw in Example 4.8–1, $\mathrm{Sh}(\mathscr{F}_{\mathsf{bdd}}) = \mathscr{G}^\nu_{\mathsf{TM}}$. Let X be a vector field possessing an integral curve $\xi : \mathbb{T} \to \mathsf{M}$ for which

$$\limsup_{t \to \sup \mathbb{T}} \|\xi'(t)\|_{\mathbb{G}} = \infty$$

(this requires that $\mathbb{T}$ be noncompact, of course).

Now let us see how this gives rise to a trajectory for $\mathrm{Sh}(\mathfrak{G}_{\mathsf{bdd}})$ that is not a trajectory for $\mathfrak{G}_{\mathsf{bdd}}$. We let $\mathbb{T}$ be the interval of definition of the integral curve ξ described above. We consider the open subset $\mathsf{M} \subseteq \mathsf{M}$. We then have the open-loop system $(X, \mathbb{T}, \mathsf{M})$ specified by letting $X(t) = X$ (abusing notation), i.e., we consider a time-independent open-loop system. It is clear, then, that $\xi \in \mathrm{Traj}(\mathbb{T}, \mathsf{M}, \mathrm{Sh}(\mathfrak{G}_{\mathsf{bdd}}))$ (since $\mathrm{Sh}(\mathfrak{G}_{\mathsf{bdd}}) = (\mathsf{M}, \mathscr{G}^\nu_{\mathsf{TM}})$ as we showed in Example 4.8–1), but that ξ cannot be a trajectory for $\mathfrak{G}_{\mathsf{bdd}}$ since any vector field possessing ξ as an integral curve cannot be bounded. ○

Thus we cannot expect sheafification to generally preserve trajectories. This should be neither a surprise nor a disappointment to us. It is gratifying, however, that sheafification *does* preserve trajectories in at least one important case.

Proposition 5.18 (Trajectories are preserved by sheafification of globally generated systems). *Let $m \in \mathbb{Z}_{\geq 0}$ and $m' \in \{0, \text{lip}\}$, let $\nu \in \{m + m', \infty, \omega\}$, and let $r \in \{\infty, \omega\}$, as required. Let $\mathfrak{G} = (\mathsf{M}, \mathscr{F})$ be a globally generated C^ν-tautological control system, let $\mathbb{T}$ be a time-domain, and let $\mathscr{O}_{\mathfrak{G}}$ be an open-loop subfamily for $\mathfrak{G}$. For a locally absolutely continuous curve $\xi : \mathbb{T} \to \mathsf{M}$ the following statements are equivalent:*

(i) $\xi \in \mathrm{Traj}(\mathbb{T}, \mathcal{U}, \mathscr{O}_{\mathfrak{G}})$ *for some open set* $\mathcal{U} \subseteq \mathsf{M}$*;*
(ii) $\xi \in \mathrm{Traj}(\mathbb{T}, \mathcal{U}', \mathrm{Sh}(\mathscr{O}_{\mathfrak{G}}))$ *for some open set* $\mathcal{U}' \subseteq \mathsf{M}$.

Proof Since $\mathscr{O}_{\mathfrak{G}}(\mathbb{T}, \mathcal{U}) \subseteq \mathrm{Sh}(\mathscr{O}_{\mathfrak{G}})(\mathbb{T}, \mathcal{U})$, the first assertion clearly implies the second. So it is the opposite implication that we need to prove.

Thus let $\mathcal{U}' \subseteq \mathsf{M}$ be open and suppose that $\xi \in \mathrm{Traj}(\mathbb{T}, \mathcal{U}', \mathrm{Sh}(\mathscr{O}_{\mathfrak{G}}))$. Let $X \in \mathrm{LI}\Gamma^\nu(\mathbb{T}; \mathsf{T}\mathcal{U}')$ be such that ξ is an integral curve for X and such that $X_t \in \mathrm{Sh}(\mathscr{F})(\mathcal{U}')$ for every $t \in \mathbb{T}$. For each *fixed* $\tau \in \mathbb{T}$, there exists $X_\tau \in \mathrm{LI}\Gamma^\nu(\mathbb{T}; \mathscr{F}(\mathsf{M}))$ such that $[X_{\tau,t}]_{\xi(\tau)} = [X_t]_{\xi(\tau)}$ for every $t \in \mathbb{T}$. (This is the definition of $\mathrm{Sh}(\mathscr{O}_{\mathfrak{G}})$, noting that $\mathscr{F}$ is globally generated.) This means that around τ we have a bounded open interval $\mathbb{T}_\tau \subseteq \mathbb{T}$ and a neighbourhood $\mathcal{U}_\tau$ of $\xi(\tau)$ so that $\xi(\mathbb{T}_\tau) \subseteq \mathcal{U}_\tau$ and so that $\xi'(t) = X_\tau(t, \xi(t))$ for almost every $t \in \mathbb{T}_\tau$. By paracompactness, we can choose a locally finite refinement of these intervals that also covers $\mathbb{T}$. By repartitioning, we arrive at a locally finite pairwise disjoint covering $(\mathbb{T}_j)_{j \in J}$ of $\mathbb{T}$ by subintervals with the following property: the index set J is a finite or countable subset of $\mathbb{Z}$ chosen so that $t_1 < t_2$ whenever $t_1 \in \mathbb{T}_{j_1}$ and $t_2 \in \mathbb{T}_{j_2}$ with $j_1 < j_2$. That is, we order the labels for the elements of the partition in the natural way, this making sense since the cover is locally finite. By construction, we have $X_j \in \mathrm{LI}\Gamma^\nu(\mathbb{T}_j; \mathscr{F}(\mathsf{M}))$ with the property that $\xi|\mathbb{T}_j$ is an integral curve for X_j. We then define $\overline{X}: \mathbb{T} \to \mathscr{F}(\mathsf{M})$ by asking that $\overline{X}|\mathbb{T}_j = X_j$. It remains to show that $\overline{X} \in \mathrm{LI}\Gamma^\nu(\mathbb{T}; \mathscr{F}(\mathsf{M}))$.

Because each of the vector fields X_j, $j \in J$, is a Carathéodory vector field, we easily conclude that $\overline{X}$ is also a Carathéodory vector field.

Let $K \subseteq \mathsf{M}$ be compact, $k \in \mathbb{Z}_{\geq 0}$, and $\boldsymbol{a} \in c_0(\mathbb{Z}_{\geq 0}; \mathbb{R}_{>0})$, and denote

$$p_K = \begin{cases} p^\infty_{K,k}, & \nu = \infty, \\ p^m_K, & \nu = m, \\ p^{m+\mathrm{lip}}_K, & \nu = m + \mathrm{lip}, \\ p^\omega_{K,\boldsymbol{a}}, & \nu = \omega. \end{cases}$$

By Lemma 3.10, for each $j \in J$, there then exists $g_j \in \mathrm{L}^1_{\mathrm{loc}}(\mathbb{T}_j; \mathbb{R}_{\geq 0})$ such that

$$p_K(X_{j,t}) \leq g_j(t), \qquad t \in \mathbb{T}_j.$$

Define $g: \mathbb{T} \to \mathbb{R}_{\geq 0}$ by asking that $g|\mathbb{T}_j = g_j$. We claim that $g \in \mathrm{L}^1_{\mathrm{loc}}(\mathbb{T}; \mathbb{R}_{\geq 0})$. Let $\mathbb{T}' \subseteq \mathbb{T}$ be a compact subinterval. The set

$$J_{\mathbb{T}'} = \{ j \in J \mid \mathbb{T}' \cap \mathbb{T}_j \neq \emptyset \}.$$

is finite by local finiteness of the cover $(\mathbb{T}_j)_{j \in J}$. Now we have

$$\int_{\mathbb{T}'} g(t)\, \mathrm{d}t \leq \sum_{j \in J_{\mathbb{T}'}} \int_{\mathbb{T}_j} g_j(t)\, \mathrm{d}t < \infty.$$

Since

$$p_K(\overline{X}_t) \leq g(t), \qquad t \in \mathbb{T},$$

from Lemma 3.10 we conclude that $\overline{X} \in \mathrm{LI}\Gamma^\nu(\mathbb{T}; \mathsf{TM})$, as desired. □

5.4 Attributes that can be Given to Tautological Control Systems

In this section we show that some typical assumptions that are made for control systems also can be made for tautological control systems. None of this is particularly earth-shattering, but it does serves as a plausibility check for our framework, letting us know that it has some common ground with familiar constructions from control theory.

A construction that often occurs in control theory is to determine a trajectory as the limit of a sequence of trajectories in some manner. To ensure the existence of such limits, the following property for tautological control systems is useful.

Definition 5.19 (Closed tautological control system). Let $m \in \mathbb{Z}_{\geq 0}$ and $m' \in \{0, \mathrm{lip}\}$, let $\nu \in \{m + m', \infty, \omega\}$, and let $r \in \{\infty, \omega\}$, as required. A C^ν-tautological control system $\mathfrak{G} = (\mathsf{M}, \mathscr{F})$ is ***closed*** if $\mathscr{F}(\mathcal{U})$ is closed in the topology of $\Gamma^\nu(\mathsf{T}\mathcal{U})$ for every open set $\mathcal{U} \subseteq \mathsf{M}$. ○

Here are some examples of control systems that give rise to closed tautological control systems.

Proposition 5.20 (Control systems with closed tautological control systems). *Let $m \in \mathbb{Z}_{\geq 0}$ and $m' \in \{0, \mathrm{lip}\}$, let $\nu \in \{m+m', \infty, \omega\}$, and let $r \in \{\infty, \omega\}$, as required. Let $\Sigma = (\mathsf{M}, F, \mathcal{C})$ be a C^ν-control system with $\mathfrak{G}_\Sigma$ the associated C^ν-tautological control system as in Example 5.2–1. Then $\mathfrak{G}_\Sigma$ is closed if Σ has either of the following two attributes:*

(i) $\mathcal{C}$ is compact;
(ii) $\mathcal{C}$ is a closed subset of $\mathbb{R}^k$ and the system is control-affine, i.e.,

$$F(x, \mathbf{u}) = f_0(x) + \sum_{a=1}^{k} u^a f_a(x),$$

for $f_0, f_1, \ldots, f_k \in \Gamma^\nu(\mathsf{TM})$.

Proof (i) Let $\mathcal{U} \subseteq \mathsf{M}$ be open. The map

$$\mathcal{C} \ni u \mapsto F^u \in \Gamma^\nu(\mathsf{T}\mathcal{U})$$

is continuous. Now let $\mathcal{U} \subseteq \mathsf{M}$ be open and note that $\mathscr{F}_\Sigma(\mathcal{U})$ is the image of $\mathcal{C}$ under the mapping

$$\mathcal{C} \ni u \mapsto F^u|\mathcal{U} \in \Gamma^\nu(\mathsf{T}\mathcal{U}).$$

Thus $\mathscr{F}_\Sigma(\mathcal{U})$ is compact, and so closed, being the image of a compact set under a continuous mapping [23, Theorem 17.7].

(ii) Let $\mathcal{U} \subseteq \mathsf{M}$ be open. Just as in the preceding part of the proof, we consider the mapping $\mathbf{u} \mapsto F^{\mathbf{u}}|\mathcal{U}$. Note that the image of the mapping

$$\mathbf{u} \mapsto F^{\mathbf{u}} = f_0 + \sum_{a=1}^{k} u^a f_a$$

is a finite-dimensional affine subspace of the $\mathbb{R}$-vector space $\Gamma^\nu(\mathsf{T}\mathcal{U})$. Therefore, this image is closed since (1) locally convex topologies are translation-invariant (by construction) and since (2) finite-dimensional subspaces of locally convex spaces are closed [13, Proposition 2.10.1]. Moreover, the map $\mathbf{u} \mapsto F^{\mathbf{u}}|\mathcal{U}$ is closed onto its image since any surjective linear map between finite-dimensional locally convex spaces is closed. We conclude, therefore, that if we restrict this map from all of $\mathbb{R}^k$ to $\mathcal{C}$, then the image is closed. □

Let us next turn to attributes of tautological control systems arising from the fact, shown in Example 5.2–4, that tautological control systems give rise to differential inclusions in a natural way.

Proposition 5.21 (Continuity of differential inclusions arising from tautological control systems). *Let $m \in \mathbb{Z}_{\geq 0}$ and $m' \in \{0, \mathrm{lip}\}$, let $\nu \in \{m + m', \infty, \omega\}$, and let $r \in \{\infty, \omega\}$, as required. If $\mathfrak{G} = (\mathsf{M}, \mathscr{F})$ is a C^ν-tautological control system, then*

(i) $\mathscr{X}_{\mathfrak{G}}$ is lower semicontinuous and
(ii) $\mathscr{X}_{\mathfrak{G}}$ is upper semicontinuous if $\mathfrak{G}$ is globally generated and $\mathscr{F}(\mathsf{M})$ is compact.

Proof (i) Let $x_0 \in \mathsf{M}$ and let $v_{x_0} \in \mathscr{X}_{\mathfrak{G}}(x_0)$. Then there exists a neighbourhood $\mathcal{W}$ of x_0 and $X \in \mathscr{F}(\mathcal{W})$ such that $X(x_0) = v_{x_0}$. Let $\mathcal{V} \subseteq \mathsf{TM}$ be a neighbourhood of v_{x_0}. By continuity of X, there exists a neighbourhood $\mathcal{U} \subseteq \mathcal{W}$ of x_0 such that $X(\mathcal{U}) \subseteq \mathcal{V}$. This implies that $X(x) \in \mathscr{X}_{\mathfrak{G}}(x)$ for every $x \in \mathcal{U}$, giving lower semicontinuity of $\mathscr{X}_{\mathfrak{G}}$.

(ii) Let $x_0 \in \mathsf{M}$ and let $\mathcal{V} \subseteq \mathsf{TM}$ be a neighbourhood of $\mathscr{X}_{\mathfrak{G}}(x_0)$. For each $X \in \mathscr{F}(\mathsf{M})$, $\mathcal{V}$ is a neighbourhood of $X(x_0)$ and so there exist neighbourhoods $\mathcal{M}_X \subseteq \mathsf{M}$ of x_0 and $\mathcal{C}_X \subseteq \mathscr{F}(\mathsf{M})$ of X such that

$$\{X'(x) \mid x \in \mathcal{M}_X,\ X' \in \mathcal{C}_X\} \subseteq \mathcal{V}.$$

Since $\mathscr{F}(\mathsf{M})$ is compact, let $X_1, \ldots, X_k \in \mathscr{F}(\mathsf{M})$ be such that $\mathscr{F}(\mathsf{M}) = \cup_{j=1}^k \mathcal{C}_{X_j}$. Then the neighbourhood $\mathcal{U} = \cap_{j=1}^k \mathcal{M}_{X_j}$ of x_0 has the property that $\mathscr{X}_{\mathfrak{G}}(\mathcal{U}) \subseteq \mathcal{V}$. □

There are many easy examples to illustrate that compactness of $\mathscr{F}(\mathsf{M})$ is generally required in part (ii) of the preceding result. Here is one.

Example 5.22 (A tautological control system with non-upper semicontinuous differential inclusion). Let $m \in \mathbb{Z}_{\geq 0}$ and $m' \in \{0, \mathrm{lip}\}$, let $\nu \in \{m + m', \infty, \omega\}$, and let $r \in \{\infty, \omega\}$, as required. Let M be a C^r-manifold and let $x_0 \in \mathsf{M}$. Let $\mathscr{F}(x_0)$ be the globally generated sheaf of sets of C^ν-vector fields defined by

$$\mathscr{F}(x_0)(\mathsf{M}) = \{X \in \Gamma^\nu(\mathsf{TM}) \mid X(x_0) = 0\}.$$

We claim that, if we take $\mathfrak{G} = (\mathsf{M}, \mathscr{F}(x_0))$, then we have

$$\mathscr{X}_{\mathfrak{G}}(x) = \begin{cases} \{0_{x_0}\}, & x = x_0, \\ \mathsf{T}_x\mathsf{M}, & x \neq x_0. \end{cases} \tag{5.1}$$

In the case $\nu = \infty$ or $\nu = m$, this is straightforward. Let $\mathcal{U}$ be a neighbourhood of $x \neq x_0$ such that $x_0 \notin \mathrm{cl}(\mathcal{U})$. By the smooth Tietze Extension Theorem [1, Proposition 5.5.8], if $X \in \Gamma^\infty(\mathsf{TM})$, then there exists $X' \in \Gamma^\infty(\mathsf{TM})$ such that $X'|\mathcal{U} = X|\mathcal{U}$ and such that $X'(x_0) = 0_{x_0}$. Thus $[X]_x = [X'_x]$ and so we have $\mathscr{F}(x_0)_x = \mathscr{G}^\nu_{x,\mathsf{M}}$ in this case. From this, (5.1) follows.

The case of $\nu = m + \mathrm{lip}$ follows as does the case $\nu = m$, noting that a locally Lipschitz vector field multiplied by a smooth function is still a locally Lipschitz vector field [22, Proposition 1.5.3].

The case of $\nu = \omega$ is a little more difficult, and relies on Cartan's Theorem A for coherent sheaves on real analytic manifolds [5]. Here is the argument for those who know a little about sheaves. First, define a sheaf of sets (in fact, submodules) of real analytic vector fields by

$$\mathscr{I}_{x_0}(\mathcal{U}) = \begin{cases} \{X \in \Gamma^\omega(\mathsf{T}\mathcal{U}) \mid X(x_0) = 0_{x_0}\}, & x_0 \in \mathcal{U}, \\ \Gamma^\omega(\mathsf{T}\mathcal{U}), & x_0 \notin \mathcal{U}. \end{cases}$$

We note that $\mathscr{I}_{x_0}$ is a coherent sheaf since it is a finitely generated subsheaf of the coherent sheaf $\mathscr{G}^\omega_{\mathsf{TM}}$ [8, Theorem 3.16].[1] Let $x \neq x_0$ and let $v_x \in \mathsf{T}_x\mathsf{M}$. By Cartan's Theorem A, there exist $X_1, \dots, X_k \in \mathscr{I}_{x_0}(\mathsf{M}) = \mathscr{F}(x_0)(\mathsf{M})$ such that $[X_1]_x, \dots, [X_k]_x$ generate $(\mathscr{I}_{x_0})_x = \mathscr{G}^\omega_{x,\mathsf{TM}}$ as a module over the ring $\mathscr{C}^\omega_{x,\mathsf{M}}$ of germs of functions at x. Let $[X]_x \in \mathscr{G}^\omega_{x,\mathsf{TM}}$ be such that $X(x) = v_x$. There then exist $[f^1]_x, \dots, [f^k]_x \in \mathscr{C}^\omega_{x,\mathsf{M}}$ such that

$$[f^1]_x[X_1]_x + \cdots + [f^k]_x[X_k]_x = [X]_x.$$

Therefore,

$$v_x = X(x) = f^1(x)X_1(x) + \cdots + f^k(x)X_k(x),$$

and so, taking

$$X = f^1X_1 + \cdots + f^kX_k \in \mathscr{I}_{x_0}(\mathsf{M}) = \mathscr{F}(x_0)(\mathsf{M}),$$

we see that $v_x = X(x) \in \mathscr{X}_{\mathfrak{G}}(x)$, which establishes (5.1) in this case.

In any event, (5.1) holds, and it is easy to see that this differential inclusion is not upper semicontinuous. ○

We can make the following definitions, rather analogous to those of Definition 3.29 for differential inclusions.

Definition 5.23 (Attributes of tautological control systems coming from the associated differential inclusion). Let $m \in \mathbb{Z}_{\geq 0}$ and $m' \in \{0, \mathrm{lip}\}$, let $\nu \in$

---

[1] This relies on the fact that Oka's Theorem, in the version of "the sheaf of sections of a vector bundle is coherent", holds in the real analytic case. It does, and the proof is the same as for the holomorphic case [8, Theorem 3.19] since the essential ingredient is the Weierstrass Preparation Theorem, which holds in the real analytic case [17, Theorem 6.1.3].

$\{m+m', \infty, \omega\}$, and let $r \in \{\infty, \omega\}$, as required. The C^ν-tautological control system $\mathfrak{G} = (\mathsf{M}, \mathscr{F})$ is:

(i) ***closed-valued*** (resp. ***compact-valued***, ***convex-valued***) at $x \in \mathsf{M}$ if $\mathscr{X}_{\mathfrak{G}}(x)$ is closed (resp., compact, convex);
(ii) ***closed-valued*** (resp. ***compact-valued***, ***convex-valued***) if $\mathscr{X}_{\mathfrak{G}}(x)$ is closed (resp., compact, convex) for every $x \in \mathsf{M}$. ∘

One can now talk about taking "hulls" under various properties. Let us discuss this for the properties of closedness and convexity. First we need the definitions we will use.

Definition 5.24 (Convex hull, closure of a tautological control system). Let $m \in \mathbb{Z}_{\geq 0}$ and $m' \in \{0, \mathrm{lip}\}$, let $\nu \in \{m+m', \infty, \omega\}$, and let $r \in \{\infty, \omega\}$, as required. Let $\mathfrak{G} = (\mathsf{M}, \mathscr{F})$ be a C^ν-tautological control system.

(i) The ***convex hull*** of $\mathfrak{G}$ is the C^ν-tautological control system $\mathrm{conv}(\mathfrak{G}) = (\mathsf{M}, \mathrm{conv}(\mathscr{F}))$, where $\mathrm{conv}(\mathscr{F})$ is the presheaf of subsets of C^ν-vector fields given by
$$\mathrm{conv}(\mathscr{F})(\mathcal{U}) = \mathrm{conv}(\mathscr{F}(\mathcal{U})),$$
the convex hull on the right being that in the $\mathbb{R}$-vector space $\Gamma^\nu(\mathsf{T}\mathcal{U})$.
(ii) The ***closure*** of $\mathfrak{G}$ is the C^ν-tautological control system
$$\mathrm{cl}(\mathfrak{G}) = (\mathsf{M}, \mathrm{cl}(\mathscr{F})),$$
where $\mathrm{cl}(\mathscr{F})$ is the presheaf of subsets of C^ν-vector fields given by $\mathrm{cl}(\mathscr{F})(\mathcal{U}) = \mathrm{cl}(\mathscr{F}(\mathcal{U}))$, the closure on the right being that in the $\mathbb{R}$-topological vector space $\Gamma^\nu(\mathsf{T}\mathcal{U})$. ∘

The reader should verify that $\mathrm{cl}(\mathscr{F})$ is indeed a presheaf.

Let us now relate the two different sorts of "hulls" we have.

Proposition 5.25 (Convex hull and closure commute with taking differential inclusions). *Let $m \in \mathbb{Z}_{\geq 0}$ and $m' \in \{0, \mathrm{lip}\}$, let $\nu \in \{m+m', \infty, \omega\}$, and let $r \in \{\infty, \omega\}$, as required. Let $\mathfrak{G} = (\mathsf{M}, \mathscr{F})$ be a C^ν-tautological control system with $\mathscr{X}_{\mathfrak{G}}$ the associated differential inclusion. Then the following statements hold:*

(i) $\mathrm{conv}(\mathscr{X}_{\mathfrak{G}}) = \mathscr{X}_{\mathrm{conv}(\mathfrak{G})}$;
(ii) $\mathscr{X}_{\mathrm{cl}(\mathfrak{G})} \subseteq \mathrm{cl}(\mathscr{X}_{\mathfrak{G}})$ *and* $\mathscr{X}_{\mathrm{cl}(\mathfrak{G})} = \mathrm{cl}(\mathscr{X}_{\mathfrak{G}})$ *if $\mathfrak{G}$ is globally generated and $\mathscr{F}(\mathsf{M})$ is bounded in the compact bornology (or, equivalently, the von Neumann bornology if $\nu \in \{\infty, \omega\}$).*

Proof (i) Let $x \in \mathsf{M}$. If $v \in \mathrm{conv}(\mathscr{X}_{\mathfrak{G}}(x))$, then there exist $v_1, \dots, v_k \in \mathscr{X}_{\mathfrak{G}}(x)$ and $c_1, \dots, c_k \in [0, 1]$ satisfying $\sum_{j=1}^k c_j = 1$ such that
$$v = c_1 v_1 + \cdots + c_k v_k.$$

Let $\mathcal{U}_1, \dots, \mathcal{U}_k$ be neighbourhoods of x and let $X_j \in \mathscr{F}(\mathcal{U}_j)$ be such that $X_j(x) = v_j$, $j \in \{1, \dots, k\}$. Then, taking $\mathcal{U} = \cap_{j=1}^k \mathcal{U}_j$,

$$c_1 X_1|\mathcal{U} + \cdots + c_k X_k|\mathcal{U} \in \operatorname{conv}(\mathscr{F}(\mathcal{U})),$$

showing that $\operatorname{conv}(\mathscr{X}_{\mathfrak{G}}(x)) \subseteq \mathscr{X}_{\operatorname{conv}(\mathfrak{G})}(x)$.

Conversely, let $v \in \mathscr{X}_{\operatorname{conv}(\mathfrak{G})}$, let $\mathcal{U}$ be a neighbourhood of x, and let $X \in \operatorname{conv}(\mathscr{F}(\mathcal{U}))$ be such that $X(x) = v$. Then

$$X = c_1 X_1 + \cdots + c_k X_k$$

for $X_1, \dots, X_k \in \mathscr{F}(\mathcal{U})$ and for $c_1, \dots, c_k \in [0, 1]$ satisfying $\sum_{j=1}^k c_j = 1$. We then have

$$v = c_1 X_1(x) + \cdots + c_k X_k(x) \in \operatorname{conv}(\mathscr{X}_{\mathfrak{G}})(x),$$

completing the proof of the proposition as concerns convex hulls.

(ii) Let $x \in \mathsf{M}$, let $v \in \mathscr{X}_{\operatorname{cl}(\mathfrak{G})}(x)$, let $\mathcal{U}$ be a neighbourhood of x, and let $X \in \operatorname{cl}(\mathscr{F}(\mathcal{U}))$ be such that $X(x) = v$. Let $(I, \preceq)$ be a directed set and let $(X_i)_{i \in I}$ be a net in $\mathscr{F}(\mathcal{U})$ converging to X in the appropriate topology. Then we have $\lim_{i \in I} X_i(x) = X(x)$ since the net $(X_i)_{i \in I}$ converges uniformly in some neighbourhood of x (this is true for all cases of ν). Thus $v \in \operatorname{cl}(\mathscr{X}_{\mathfrak{G}}(x))$, as desired.

Suppose that $\mathscr{F}$ is globally generated with $\mathscr{F}(\mathsf{M})$ bounded, let $x \in \mathsf{M}$, and let $v \in \operatorname{cl}(\mathscr{X}_{\mathfrak{G}})(x)$. Thus there exists a sequence $(v_j)_{j \in \mathbb{Z}_{>0}}$ in $\mathscr{X}_{\mathfrak{G}}(x)$ converging to v. Let $X_j \in \mathscr{F}(\mathsf{M})$ be such that $X_j(x) = v_j$, $j \in \mathbb{Z}_{>0}$. Since $\operatorname{cl}(\mathscr{F}(\mathsf{M}))$ is compact, there is a subsequence $(X_{j_k})_{j_k}$ in $\mathscr{F}(\mathsf{M})$ converging to $X \in \operatorname{cl}(\mathscr{F}(\mathsf{M}))$. Moreover,

$$X(x) = \lim_{k \to \infty} X_{j_k}(x) = \lim_{j \to \infty} v_j = v$$

since $(X_{j_k})_{k \in \mathbb{Z}_{>0}}$ converges to X uniformly in some neighbourhood of x (again, this is true for all ν). Thus $v \in \mathscr{X}_{\operatorname{cl}(\mathfrak{G})}(x)$.

The parenthetical comment in the final assertion of the proof follows since the compact and von Neumann bornologies agree for nuclear spaces [19, Proposition 4.47]. □

The following example shows that the opposite inclusion stated in the proposition for closures does not generally hold.

Example 5.26 (A tautological control system for which the closure of the differential inclusion is not the differential inclusion of the closure). We will talk our way through a general sort of example, leaving to the reader the job of instantiating this to give a concrete example.

Let $m \in \mathbb{Z}_{>0}$ and $m' \in \{0, \mathrm{lip}\}$, let $\nu \in \{m + m', \infty, \omega\}$, and let $r \in \{\infty, \omega\}$, as required. Let M be a C^r-manifold. Let $x \in \mathsf{M}$ and let $(X_j)_{j \in \mathbb{Z}_{>0}}$ be a sequence of C^ν-vector fields with the following properties:

1. $(X_j(x))_{j \in \mathbb{Z}_{>0}}$ converges to 0_x;

2. $X_j(x) \neq 0_x$ for all $j \in \mathbb{Z}_{>0}$;
3. there exists a neighbourhood $\mathcal{O}$ of zero in $\Gamma^\nu(\mathsf{TM})$ such that, for each $j \in \mathbb{Z}_{>0}$,

$$\{k \in \mathbb{Z}_{>0} \setminus \{j\} \mid X_k - X_j \in \mathcal{O}\} = \emptyset.$$

Let $\mathscr{F}$ be the globally generated presheaf of sets of C^ν-vector fields given by $\mathscr{F}(\mathsf{M}) = \{X_j \mid j \in \mathbb{Z}_{>0}\}$. Then $0_x \in \mathrm{cl}(\mathscr{X}_{\mathfrak{G}}(x))$. We claim that $0_x \notin \mathscr{X}_{\mathrm{cl}(\mathfrak{G})}(x)$. To see this, suppose that $0_x \in \mathscr{X}_{\mathrm{cl}(\mathfrak{G})}(x)$. Since $\mathscr{F}(\mathsf{M})$ is countable, this implies that there is a subsequence $(X_{j_k})_{k\in\mathbb{Z}_{>0}}$ that converges in $\Gamma^\nu(\mathsf{TM})$. But this is prohibited by the construction of the sequence $(X_j)_{j\in\mathbb{Z}_{>0}}$. ∘

5.5 Trajectory Correspondences with Other Sorts of Control Systems

In Example 5.2 and Proposition 5.3 we made precise the connections between various models for control systems: control systems, differential inclusions, and tautological control systems. In order to flesh out these connections more deeply, in this section we investigate the possible correspondences between the trajectories for the various models.

We first consider correspondences between trajectories of control systems and their associated tautological control systems. Thus we let $m \in \mathbb{Z}_{\geq 0}$ and $m' \in \{0, \mathrm{lip}\}$, let $\nu \in \{m + m', \infty, \omega\}$, and let $r \in \{\infty, \omega\}$, as required. Let $\Sigma = (\mathsf{M}, F, \mathcal{C})$ be a C^ν-control system with $\mathfrak{G}_\Sigma$ the associated C^ν-tautological control system, as in Example 5.2–1. As we saw in Proposition 5.3(ii), the correspondence between Σ and $\mathfrak{G}_\Sigma$ is perfect, at the system level, when the map $u \mapsto F^u$ is injective and open onto its image. Part (ii) of the following result shows that this perfect correspondence almost carries over at the level of trajectories as well. Included with this statement we include a few other related ideas concerning trajectory correspondences.

Theorem 5.27 (Correspondence between trajectories of a control system and its associated tautological control system). *Let $m \in \mathbb{Z}_{\geq 0}$ and $m' \in \{0, \mathrm{lip}\}$, let $\nu \in \{m + m', \infty, \omega\}$, and let $r \in \{\infty, \omega\}$, as required. Let $\Sigma = (\mathsf{M}, F, \mathcal{C})$ be a C^ν-control system with $\mathfrak{G}_\Sigma$ the associated C^ν-tautological control system, as in Example 5.2–1. Then the following statements hold:*

(i) $\mathrm{Traj}(\mathbb{T}, \mathcal{U}, \Sigma) \subseteq \mathrm{Traj}(\mathbb{T}, \mathcal{U}, \mathscr{O}_{\mathfrak{G}_\Sigma,\mathrm{cpt}})$;
(ii) if the map $u \mapsto F^u$ is injective and proper, then $\mathrm{Traj}(\mathbb{T}, \mathcal{U}, \mathscr{O}_{\mathfrak{G}_\Sigma,\mathrm{cpt}}) \subseteq \mathrm{Traj}(\mathbb{T}, \mathcal{U}, \Sigma)$;
(iii) if $\mathcal{C}$ is a Suslin topological space[2] *and if F is proper, then* $\mathrm{Traj}(\mathbb{T}, \mathcal{U}, \mathscr{O}_{\mathfrak{G}_\Sigma,\mathrm{cpt}}) \subseteq \mathrm{Traj}(\mathbb{T}, \mathcal{U}, \Sigma)$.

[2] Recall that this means that $\mathcal{C}$ is the continuous image of a complete, separable, metric space. We refer to [3, §6.6–6.8] for an outline of the theory of Suslin spaces.

(iv) if, in addition, $\nu \in \{\infty, \omega\}$, then we may replace $\mathrm{Traj}(\mathbb{T}, \mathcal{U}, \mathscr{O}_{\mathfrak{G}_\Sigma,\mathrm{cpt}})$ *with* $\mathrm{Traj}(\mathbb{T}, \mathcal{U}, \mathscr{O}_{\mathfrak{G}_\Sigma,\infty})$ *in statements (i), (ii), and (iii).*

Proof (i) Let $\xi \in \mathrm{Traj}(\mathbb{T}, \mathcal{U}, \Sigma)$ and let $\mu \in \mathrm{L}^{\mathrm{cpt}}_{\mathrm{loc}}(\mathbb{T}; \mathcal{C})$ be such that

$$\xi'(t) = F(\xi(t), \mu(t)), \qquad \text{a.e. } t \in \mathbb{T}.$$

Note that, as we saw in Example 5.2, $F^\mu|\mathcal{U} \in \mathscr{O}_{\mathfrak{G}_\Sigma,\mathrm{cpt}}(\mathbb{T}, \mathcal{U})$, making sure to note that the conclusions of Proposition 3.20 imply that $F^\mu \in \mathrm{LB}\Gamma^\nu(\mathbb{T}; \mathsf{TM})$. Thus $\xi \in \mathrm{Traj}(\mathbb{T}, \mathcal{U}, \mathscr{O}_{\mathfrak{G}_\Sigma,\infty})$. To show that, in fact, $\xi \in \mathrm{Traj}(\mathbb{T}, \mathcal{U}, \mathscr{O}_{\mathfrak{G}_\Sigma,\mathrm{cpt}})$, let $\mathbb{T}' \subseteq \mathbb{T}$ be a compact subinterval and let $K \subseteq \mathcal{C}$ be a compact set such that $\mu(t) \in K$ for almost every $t \in \mathbb{T}'$. Denote

$$\begin{aligned} \hat{F} \colon \mathcal{C} &\to \Gamma^\nu(\mathsf{TM}) \\ u &\mapsto F^u. \end{aligned}$$

Since $\hat{F}$ is continuous, $\hat{F}(K)$ is compact [23, Theorem 17.7]. Since $F^\mu_t \in \hat{F}(K)$ for almost every $t \in \mathbb{T}'$, we conclude that $\xi \in \mathrm{Traj}(\mathbb{T}, \mathcal{U}, \mathscr{O}_{\mathfrak{G}_\Sigma,\mathrm{cpt}})$, as claimed.

(ii) Recall from [4, Proposition I.10.2] that, if $\hat{F}$ (as defined above) is proper, then it has a closed image, and is an homeomorphism onto its image. If $\xi \in \mathrm{Traj}(\mathbb{T}, \mathcal{U}, \mathscr{O}_{\mathfrak{G}_\Sigma,\mathrm{cpt}})$, then there exists $X \in \mathscr{O}_{\mathfrak{G}_\Sigma,\mathrm{cpt}}$ such that $\xi'(t) = X(t, \xi(t))$ for almost every $t \in \mathbb{T}$. Note that, since $X \in \mathscr{O}_{\mathfrak{G}_\Sigma,\mathrm{cpt}}$, we have $X(t) \in \mathscr{F}_\Sigma(\mathsf{M}) = \mathrm{image}(\hat{F})$. Thus, by hypothesis, there exists a unique $\mu \colon \mathbb{T} \to \mathcal{C}$ such that $\hat{F} \circ \mu = X$. To show that μ is measurable, let $\mathcal{O} \subseteq \mathcal{C}$ be open so that $\hat{F}(\mathcal{O})$ is an open subset of $\mathrm{image}(\hat{F})$. Thus there exists an open set $\mathcal{O}' \subseteq \Gamma^\nu(\mathsf{TM})$ such that $\hat{F}(\mathcal{O}) = \mathrm{image}(\hat{F}) \cap \mathcal{O}'$. Then we have

$$\mu^{-1}(\mathcal{O}) = X^{-1}(\hat{F}(\mathcal{O})) = X^{-1}(\mathcal{O}'),$$

giving the desired measurability. To show that $\mu \in \mathrm{L}^{\mathrm{cpt}}_{\mathrm{loc}}(\mathbb{T}; \mathcal{C})$, let $\mathbb{T}' \subseteq \mathbb{T}$ be a compact subinterval and let $K \subseteq \Gamma^\nu(\mathsf{TM})$ be such that $X(t) \in K$ for almost every $t \in \mathbb{T}'$. Then, since $\hat{F}$ is proper, $\hat{F}^{-1}(K)$ is a compact subset of $\mathcal{C}$. Since $\mu(t) \in \hat{F}^{-1}(K)$ for almost every $t \in \mathbb{T}'$ we conclude that $\mu \in \mathrm{L}^{\mathrm{cpt}}_{\mathrm{loc}}(\mathbb{T}; \mathcal{C})$.

(iii) Let $\xi \in \mathrm{Traj}(\mathbb{T}, \mathcal{U}, \mathscr{O}_{\mathfrak{G}_\Sigma,\mathrm{cpt}})$ and let $X \in \mathscr{O}_{\mathfrak{G}_\Sigma,\mathrm{cpt}}(\mathbb{T}, \mathcal{U})$ be such that $\xi'(t) = X(t, \xi(t))$ for almost every $t \in \mathbb{T}$. We wish to construct $\mu \in \mathrm{L}^{\mathrm{cpt}}_{\mathrm{loc}}(\mathbb{T}, \mathcal{C})$ such that

$$\xi' = F(\xi(t), \mu(t)), \qquad \text{a.e. } t \in \mathbb{T}.$$

We fix an arbitrary element $\bar{u} \in \mathcal{C}$ (it matters not which) and then define a set-valued map $U \colon \mathbb{T} \twoheadrightarrow \mathcal{C}$ by

$$U(t) = \begin{cases} \{u \in \mathcal{C} \mid \xi'(t) = F(\xi(t), u)\}, & \xi'(t) \text{ exists,} \\ \{\bar{u}\}, & \text{otherwise.} \end{cases}$$

Since $X(t) \in \mathscr{F}_\Sigma(\mathsf{M})$, we conclude that $X(t) \in \text{image}(\hat{F})$ for every $t \in \mathbb{T}$, i.e., $X(t) = F^u$ for some $u \in \mathcal{C}$, and so $U(t) \neq \emptyset$ for every $t \in \mathbb{T}$.

Properness of F ensures that $U(t)$ is compact for every $t \in \mathbb{T}$. The following lemma shows that *any* selection μ of U is locally essentially bounded in the compact bornology.

Lemma 1 *If* $\mathbb{T}' \subseteq \mathbb{T}$ *is a compact subinterval, then the set* $\cup\{U(t) \mid t \in \mathbb{T}'\}$ *is contained in a compact subset of* $\mathcal{C}$.

Proof Let us define $F_\xi : \mathbb{T} \times \mathcal{C} \to \mathsf{TM}$ by $F_\xi(t, u) = F(\xi(t), u)$. We claim that, if $\mathbb{T}' \subseteq \mathbb{T}$ is compact, then $F_\xi|\mathbb{T}' \times \mathcal{C}$ is proper. To see this, first define

$$\begin{aligned} G_\xi : \mathbb{T}' \times \mathcal{C} &\to \mathsf{M} \times \mathcal{C} \\ (t, u) &\mapsto (\xi(t), u), \end{aligned}$$

i.e., $G_\xi = \xi \times \mathrm{id}_\mathcal{C}$. With this notation, we have $F_\xi = F \circ G_\xi$. Since $F_\xi^{-1}(K) = G_\xi^{-1}(F^{-1}(K))$ and since F is proper, to show that F_ξ is proper it suffices to show that G_ξ is proper. Let $K \subseteq \mathsf{M} \times \mathcal{C}$ be compact. We let $\mathrm{pr}_1 : \mathsf{M} \times \mathcal{C} \to \mathsf{M}$ and $\mathrm{pr}_2 : \mathsf{M} \times \mathcal{C} \to \mathcal{C}$ be the projections. Note that

$$G_\xi^{-1}(K) = (\xi \times \mathrm{id}_\mathcal{C})^{-1}(K) \subseteq \xi^{-1}(\mathrm{pr}_1(K)) \times \mathrm{id}_\mathcal{C}^{-1}(\mathrm{pr}_2(K)).$$

Since the projections are continuous, $\mathrm{pr}_1(K)$ and $\mathrm{pr}_2(K)$ are compact [23, Theorem 17.7]. Since ξ is a continuous function whose domain (for our present purposes) is the compact set $\mathbb{T}'$, $\xi^{-1}(\mathrm{pr}_1(K))$ is compact. Since the identity map is proper, $\mathrm{id}_\mathcal{C}^{-1}(\mathrm{pr}_2(K))$ is compact. Thus $G_\xi^{-1}(K)$ is contained in a product of compact sets. Since a product of compact sets is compact [23, Theorem 17.8] and $G_\xi^{-1}(K)$ is closed by continuity of G_ξ, it follows that $G_\xi^{-1}(K)$ is compact, as claimed. Thus $F_\xi|\mathbb{T}' \times \mathcal{C}$ is proper.

Now, since ξ is a trajectory for the $\mathscr{O}_{\mathfrak{G}_\Sigma,\mathrm{cpt}}$ open-loop subfamily, there exists a compact set $K' \subseteq \mathsf{TM}$ such that

$$\{\xi'(t) \mid t \in \mathbb{T}'\} \subseteq K',$$

adopting the convention that $\xi'(t)$ is taken to satisfy $\xi'(t) = F(\xi(t), \bar{u})$ when $\xi'(t)$ does not exist; this is an arbitrary and inconsequential choice. By our argument above, $K'' \triangleq (F_\xi|\mathbb{T}' \times \mathcal{C})^{-1}(K')$ is compact. Therefore, for each $t \in \mathbb{T}'$,

$$\begin{aligned} \{(t, u) \in \mathbb{T}' \times \mathcal{C} \mid u \in U(t)\} &= \{(t, u) \in \mathbb{T}' \times \mathcal{C} \mid F(\xi(t), u) = \xi'(t)\} \\ &\subseteq \{(t, u) \in \mathbb{T}' \times \mathcal{C} \mid F(\xi(t), u) \in K'\} \subseteq K''. \end{aligned}$$

Defining the compact set (compact by [23, Theorem 17.7]) $K = \mathrm{pr}_2(K'')$, with $\mathrm{pr}_2 : \mathbb{T}' \times \mathcal{C} \to \mathcal{C}$ the projection, we then have

$$\cup\{U(t) \mid t \in \mathbb{T}'\} \subseteq K. \qquad \triangledown$$

We shall now make a series of observations about the set-valued map U, using results of [12] on measurable set-valued mappings, particularly with values in Suslin spaces.

Lemma 2 *The set-valued map U is measurable, i.e., if $\mathcal{O} \subseteq \mathcal{C}$ is open, then*

$$U^{-1}(\mathcal{O}) = \{t \in \mathbb{T} \mid U(t) \cap \mathcal{O} \neq \emptyset\}$$

is measurable.

Proof Define

$$\begin{aligned} F_\xi \colon \mathbb{T} \times \mathcal{C} &\to \mathsf{TM} \\ (t, u) &\mapsto F(\xi(t), u), \end{aligned}$$

noting that $t \mapsto F_\xi(t, u)$ is measurable for each $u \in \mathcal{C}$ and that $u \mapsto F_\xi(t, u)$ is continuous for every $t \in \mathbb{T}$. It follows from [12, Theorem 6.4] that U is measurable as stated. $\triangledown$

Lemma 3 *There exists a measurable function $\mu \colon \mathbb{T} \to \mathcal{C}$ such that $\mu(t) \in U(t)$ for almost every $t \in \mathbb{T}$.*

Proof First note that $U(t)$ is a closed subset of $\mathcal{C}$ since it is either the singleton $\{\bar{u}\}$ or the preimage of the closed set $\{\xi'(t)\}$ under the continuous map $u \mapsto F(\xi(t), u)$. It follows from [12, Theorem 3.5] that

$$\operatorname{graph}(U) = \{(t, u) \in \mathbb{T} \times \mathcal{C} \mid u \in U(t)\}$$

is measurable with respect to the product σ-algebra of the Lebesgue measurable sets in $\mathbb{T}$ and the Borel sets in $\mathcal{C}$. The lemma now follows from [12, Theorem 5.7]. $\triangledown$

Now, for $t \in \mathbb{T}$ having the property that $\xi'(t)$ exists and that $\mu(t) \in U(t)$ (with μ from the preceding lemma), we have $\xi'(t) = F(\xi(t), \mu(t))$.

(iv) This follows by our observation of Example 5.10–3. □

Let us make some comments on the hypotheses of the preceding theorem.

Remarks 5.28 (Trajectory correspondence between control systems and tautological control systems)

1. Part (ii) of the result has assumptions that the map $u \mapsto F^u$ be injective and proper. An investigation of the proof shows that injectivity and openness onto the image of this map are enough to give trajectories for Σ that correspond to measurable controls. The additional assumption of properness, which gives the further consequence of the image of the map $u \mapsto F^u$ being closed, allows us to conclude boundedness of the controls. Let us look at these assumptions.

(a) By the map $u \mapsto F^u$ being injective, we definitely do not mean that the map $u \mapsto F(x,u)$ is injective for each $x \in \mathsf{M}$; this is a very strong assumption whose adoption eliminates a large number of interesting control systems. For example, if we take $\mathsf{M} = \mathbb{R}^n$, $\mathcal{C} = \mathbb{R}^k$, and

$$F(\mathbf{x}, \mathbf{u}) = \Big(\mathbf{A} + \sum_{j=1}^{k} u^j \mathbf{B}_j\Big)\mathbf{x}$$

for $n \times n$ matrices $\mathbf{A}, \mathbf{B}_1, \dots, \mathbf{B}_k$, then we have a bilinear control system [9]. In this case, the map $\mathbf{u} \mapsto F(\mathbf{0}, \mathbf{u})$ is never injective, while the map $\mathbf{u} \mapsto F^{\mathbf{u}}$ may very well be.

(b) Let us take $\mathsf{M} = \mathbb{R}^2$, $\mathcal{C} = \mathbb{R}$, and

$$F((x_1, x_2), u) = f_1(u)\frac{\partial}{\partial x_1} + f_2(u)\frac{\partial}{\partial x_2},$$

where $f_1, f_2 \colon \mathbb{R} \to \mathbb{R}$ are such that the map $u \mapsto (f_1(u), f_2(u))$ is injective and continuous, but not an homeomorphism onto its image. Such a system may be verified to be a C^ν-control system for any $\nu \in \{m + m', \infty, \omega\}$ with $m \in \mathbb{Z}_{\geq 0}$ and $m' \in \{0, \mathrm{lip}\}$ (using Theorem 3.18). In this case, we claim that the map $\hat{F} \colon u \mapsto F^u$ is injective and continuous, but not an homeomorphism onto its image. Injectivity of the map is clear and continuity follows since F is a jointly parameterised vector field of class C^ν. Define a linear map

$$\begin{aligned} \kappa \colon \mathbb{R}^2 &\to \Gamma^\nu(\mathsf{TM}) \\ (v_1, v_2) &\mapsto v_1\frac{\partial}{\partial x_1} + v_2\frac{\partial}{\partial x_2}, \end{aligned}$$

i.e., $\kappa(\mathbf{v})$ is the constant vector field with components (v_1, v_2). Since κ has as domain a finite-dimensional vector space, it is continuous, and so is an homeomorphism onto its closed image (arguing as in the proof of Proposition 5.20(ii)). Then $\hat{F} = \kappa \circ (f_1 \times f_2)$, and so we conclude that $\hat{F}$ is an homeomorphism onto its image if and only if $f_1 \times f_2$ is an homeomorphism onto its image, and this gives our claim.

(c) Let us take $\mathsf{M} = \mathbb{R}$, $\mathcal{C} = \mathbb{R}$, and $F(x,u) = \tan^{-1}(u)\frac{\partial}{\partial x}$. As with the example above, we regard this as a control system of class C^ν for any $\nu \in \{m + m', \infty, \omega\}$, for $m \in \mathbb{Z}_{\geq 0}$ and $m' \in \{0, \mathrm{lip}\}$. We claim that $\hat{F} \colon u \mapsto F^u$ is an homeomorphism onto its image, but is not proper. This is verified in exactly the same manner as in the preceding example.

(d) If $\mathcal{C}$ is compact, then $\hat{F}$ is proper because, if $K \subseteq \Gamma^\nu(\mathsf{TM})$ is compact, then $\hat{F}^{-1}(K)$ is closed, and so compact [23, Theorem 17.5]. This gives trajectory correspondence between a C^ν-control system and its corresponding tautological control system for compact control sets when the map $\hat{F}$ is injective.

2. Part (iii) of the result has two assumptions, that $\mathcal{C}$ is a Suslin space and that F is proper. Let us consider some cases where these hypotheses hold.
   (a) Complete separable metric spaces are Suslin spaces.
   (b) If $\mathcal{C}$ is an open or a closed subspace of Suslin space, it is a Suslin space [3, Lemma 6.6.5(ii)].
   (c) For $m \in \mathbb{Z}_{\geq 0}$, $m' \in \{0, \text{lip}\}$, and $\nu \in \{m + m', \infty, \omega\}$, $\Gamma^\nu(\mathsf{TM})$ is a Suslin space. In all except the case of $\nu = \omega$, this follows since $\Gamma^\nu(\mathsf{TM})$ is a separable, complete, metrisable space. However, $\Gamma^\omega(\mathsf{TM})$ is not metrisable. Nonetheless, it is Suslin, as argued in [15, §5.3].
   (d) If $\mathcal{C}$ is compact, then F is proper. Indeed, if $K \subseteq \mathsf{TM}$ is compact, then $\pi_{\mathsf{TM}}(K)$ is compact, and
$$F^{-1}(K) \subseteq \pi_{\mathsf{TM}}(K) \times \mathcal{C},$$
and so the set on the left is compact, being a closed subset of a compact set [23, Theorem 17.5]. ○

We also have a version of the preceding theorem in the case that the control set $\mathcal{C}$ is a subset of a locally convex topological vector space, cf. Proposition 3.24. Here we also specialise for one of the implications to control-linear systems introduced in Example 3.23.

Theorem 5.29 (Correspondence between trajectories of a control-linear system and its associated tautological control system). *Let $m \in \mathbb{Z}_{\geq 0}$ and $m' \in \{0, \text{lip}\}$, let $\nu \in \{m + m', \infty, \omega\}$, and let $r \in \{\infty, \omega\}$, as required. Let $\Sigma = (\mathsf{M}, F, \mathcal{C})$ be a C^ν-sublinear control system for which $\mathcal{C}$ is a subset of a locally convex topological vector space V, and let $\mathfrak{G}_\Sigma$ be the associated C^ν-tautological control system, as in Example 5.2–1. If $\mathbb{T}$ is a time-domain and if $\mathcal{U}$ is open, then* $\mathrm{Traj}(\mathbb{T}, \mathcal{U}, \Sigma) \subseteq \mathrm{Traj}(\mathbb{T}, \mathcal{U}, \mathscr{O}_{\mathfrak{G}_\Sigma,\mathit{full}})$.

Conversely, if

(i) Σ is a C^ν-control-linear system, i.e., there exists $\Lambda \in \mathrm{L}(\mathsf{V}; \Gamma^\nu(\mathsf{TM}))$ such that $F(x, u) = \Lambda(u)(x)$,
(ii) Λ is injective, and
(iii) Λ is an open mapping onto its image,

then it is also the case that $\mathrm{Traj}(\mathbb{T}, \mathcal{U}, \mathscr{O}_{\mathfrak{G}_\Sigma,\mathit{full}}) \subseteq \mathrm{Traj}(\mathbb{T}, \mathcal{U}, \Sigma)$.

Proof We first show that $\mathrm{Traj}(\mathbb{T}, \mathcal{U}, \Sigma) \subseteq \mathrm{Traj}(\mathbb{T}, \mathcal{U}, \mathscr{O}_{\mathfrak{G}_\Sigma,\mathrm{full}})$. Suppose that $\xi \in \mathrm{Traj}(\mathbb{T}, \mathcal{U}, \Sigma)$. Thus there exists $\mu \in \mathrm{L}^1_{\mathrm{loc}}(\mathbb{T}; \mathcal{C})$ such that
$$\xi'(t) = F(\xi(t), \mu(t)), \quad \text{a.e. } t \in \mathbb{T}.$$
By Proposition 3.24 and Example 5.7, $F^\mu|\mathcal{U} \in \mathscr{O}_{\mathfrak{G}_\Sigma,\mathrm{full}}(\mathbb{T}, \mathcal{U})$ and so $\xi \in \mathrm{Traj}(\mathbb{T}, \mathcal{U}, \mathscr{O}_{\mathfrak{G}_\Sigma,\mathrm{full}})$.

Now let us prove the "conversely" assertion of the theorem. Thus we let $\xi \in \mathrm{Traj}(\mathbb{T}, \mathcal{U}, \mathscr{O}_{\mathfrak{G}_\Sigma,\mathrm{full}})$ so that there exists $X \in \mathscr{O}_{\mathfrak{G}_\Sigma,\mathrm{full}}(\mathbb{T}, \mathcal{U})$ for which $\xi'(t) = X(t, \xi(t))$ for almost every $t \in \mathbb{T}$. Since Λ is injective and since $X_t \in \Lambda(\mathcal{C})$ for each $t \in \mathbb{T}$ (this is the definition of $\mathfrak{G}_\Sigma$), we uniquely define $\mu(t) \in \mathcal{C}$ by $\Lambda(\mu(t)) = X_t$. We need only show that μ is locally Bochner integrable. Let Λ^{-1} denote the inverse of Λ, thought of as a map from image(Λ) to V. As Λ is open, Λ^{-1} is continuous. From this, measurability of μ follows immediately. To show that μ is locally Bochner integrable, let q be a continuous seminorm for the locally convex topology of V and, as per [20, §III.1.1], let p be a continuous seminorm for the locally convex topology of $\Gamma^\nu(\mathsf{TM})$ such that $q(\Lambda^{-1}(Y)) \leq p(Y)$ for every $Y \in \Gamma^\nu(\mathsf{TM})$. Then we have, for any compact subinterval $\mathbb{T}' \subseteq \mathbb{T}$,

$$\int_{\mathbb{T}'} q(\mu(t))\,\mathrm{d}t \leq \int_{\mathbb{T}'} p(X_t)\,\mathrm{d}t < \infty,$$

giving Bochner integrability of μ by Lemma 3.10. □

Let us make some observations about the preceding theorem.

Remarks 5.30 (Trajectory correspondence between control systems and tautological control systems). The converse part of Theorem 5.29 has three hypotheses: that the system is control-linear; that the map from controls to vector fields is injective; that the map from controls to vector fields is open onto its image. The first hypothesis, linearity of the system, cannot be weakened except in sort of artificial ways. As can be seen from the proof, linearity allows us to talk about the integrability of the associated control. Injectivity can be assumed without loss of generality by quotienting out the kernel if it is not. Let us consider some cases where the third hypothesis holds. Let $m \in \mathbb{Z}_{\geq 0}$ and $m' \in \{0, \mathrm{lip}\}$, let $\nu \in \{m + m', \infty, \omega\}$, and let $r \in \{\infty, \omega\}$, as required.

1. Let $\mathcal{C} \subseteq \mathbb{R}^k$ and suppose that our system is C^ν-control-affine, i.e.,

$$F(x, \mathbf{u}) = f_0(x) + \sum_{a=1}^{k} u^a f_a(x)$$

for C^ν-vector fields $f_0, f_1, \dots, f_m$. As we pointed out in Example 3.23, this can be regarded as a control-linear system by taking $\mathsf{V} = \mathbb{R} \oplus \mathbb{R}^k$

$$\mathcal{C}' = \{(u^0, \mathbf{u}) \in \mathsf{V} \mid u^0 = 1,\ \mathbf{u} \in \mathcal{C}\},$$

and

$$\Lambda(u^0, \mathbf{u}) = \sum_{a=0}^{k} u^a f_a.$$

We can assume that Λ is injective, as mentioned above. In this case, the map Λ is an homeomorphism onto its image since any map from a finite-dimensional

locally convex space is continuous [13, Proposition 2.10.2]. Thus Theorem 5.29 applies to control-affine systems, and gives trajectory equivalence in this case.
2. The other case of interest to us is that when $\mathsf{V} = \Gamma^\nu(\mathsf{TM})$ and when $\mathcal{C} \subseteq \mathsf{V}$ is then a family of globally defined vector fields of class C^ν on M. In this case, we take Λ to be the identity map on $\Gamma^\nu(\mathsf{TM})$, so the hypotheses of Theorem 5.29 are easily satisfied. The trajectory equivalence one gets in this case is that between a globally generated tautological control system and its corresponding control system as in Example 5.2–2. ○

One of the conclusions enunciated above is sufficiently interesting to justify its own theorem.

Theorem 5.31 (Correspondence between trajectories of a tautological control system and its associated control system). *Let $m \in \mathbb{Z}_{\geq 0}$ and $m' \in \{0, \mathrm{lip}\}$, let $\nu \in \{m + m', \infty, \omega\}$, and let $r \in \{\infty, \omega\}$, as required. Let $\mathfrak{G} = (\mathsf{M}, \mathscr{F})$ be a globally generated C^ν-tautological control system. As in Example 5.2–2, let $\Sigma_{\mathfrak{G}} = (\mathsf{M}, \Sigma_{\mathfrak{G}}, \mathcal{C}_{\mathscr{F}})$ be the corresponding C^ν-control system. Then, for each time-domain $\mathbb{T}$ and each open set $\mathcal{U} \subseteq \mathsf{M}$, $\mathrm{Traj}(\mathbb{T}, \mathcal{U}, \mathscr{O}_{\mathfrak{G},full}) = \mathrm{Traj}(\mathbb{T}, \mathcal{U}, \Sigma_{\mathfrak{G}})$.*

Proof This is the observation made in Remark 5.30–2. □

Now we turn to relationships between trajectories for tautological control systems and differential inclusions. In Example 5.2–3 we showed how a tautological control system can be built from a differential inclusion. However, as we mentioned in that example, we cannot expect any sort of general correspondence between trajectories of the differential inclusion and the tautological control system constructed from it; differential inclusions are just too irregular. We can, however, consider the correspondence in the other direction, as the following theorem indicates.

Theorem 5.32 (Correspondence between trajectories of a tautological control system and its associated differential inclusion). *Let $m \in \mathbb{Z}_{\geq 0}$ and $m' \in \{0, \mathrm{lip}\}$, let $\nu \in \{m + m', \infty, \omega\}$, and let $r \in \{\infty, \omega\}$, as required. Let $\mathfrak{G} = (\mathsf{M}, \mathscr{F})$ be a C^ν-tautological control system and let $\mathscr{X}_{\mathfrak{G}}$ be the associated differential inclusion, as in Example 5.2–4. For $\mathbb{T}$ a time-domain and $\mathcal{U} \subseteq \mathsf{M}$ an open set, $\mathrm{Traj}(\mathbb{T}, \mathcal{U}, \mathfrak{G}) \subseteq \mathrm{Traj}(\mathbb{T}, \mathcal{U}, \mathscr{X}_{\mathfrak{G}})$.*

Conversely, if $\mathscr{F}$ is globally generated and if $\mathscr{F}(\mathsf{M})$ is a compact subset of $\Gamma^\nu(\mathsf{TM})$, then $\mathrm{Traj}(\mathbb{T}, \mathcal{U}, \mathscr{X}_{\mathfrak{G}}) \subseteq \mathrm{Traj}(\mathbb{T}, \mathcal{U}, \mathfrak{G})$.

Proof Since, for an open-loop system $(X, \mathbb{T}, \mathcal{U})$, $X(t) \in \mathscr{F}(\mathcal{U})$ for every $t \in \mathbb{T}$, we have $X(t, x) \in \mathscr{X}_{\mathfrak{G}}(x)$ for every $(t, x) \in \mathbb{T} \times \mathcal{U}$. Thus, if $\xi \in \mathrm{Traj}(\mathbb{T}, \mathcal{U}, \mathfrak{G})$, then we have $\xi'(t) \in \mathscr{X}_{\mathfrak{G}}(\xi(t))$ for almost every $t \in \mathbb{T}$.

For the "conversely" part of the theorem, if ξ is a trajectory for the differential inclusion $\mathscr{X}_{\mathfrak{G}}$ then, for almost every $t \in \mathbb{T}$, $\xi'(t) = X(\xi(t))$ for some $X \in \mathscr{F}(\mathsf{M})$. Therefore, let us fix an arbitrary $\overline{X} \in \mathscr{F}(\mathsf{M})$ and let us define $U : \mathbb{T} \twoheadrightarrow \mathscr{F}(\mathsf{M})$ by

$$U(t) = \begin{cases} \{X \in \mathscr{F}(\mathsf{M}) \mid \xi'(t) = X(\xi(t))\}, & \xi'(t) \text{ exists}, \\ \{\overline{X}\}, & \text{otherwise}. \end{cases}$$

Now we note that

1. $\mathcal{C}_{\mathscr{F}} = \mathscr{F}(\mathsf{M})$ is a Suslin space, being a closed subset of a Suslin space, and
2. the map $F_{\mathscr{F}}$ is proper by Remark 5.28–2d.

Thus we are in exactly the right framework to use the proof of Theorem 5.27(iii) to show that there exists a locally essentially bounded (in the compact bornology) measurable control $t \mapsto X(t)$ for which

$$\xi'(t) = F_{\mathscr{F}}(\xi(t), X(t)), \qquad \text{a.e. } t \in \mathbb{T},$$

and so $\xi \in \mathrm{Traj}(\mathbb{T}, \mathfrak{U}, \Sigma_{\mathfrak{G}})$, as desired. □

Let us comment on the hypotheses of this theorem.

Remark 5.33 (Trajectory correspondence between tautological control systems and differential inclusions). The assumption that $\mathscr{F}(\mathsf{M})$ be compact in the "conversely" part of the preceding theorem is indispensable. The connection going from differential inclusion to tautological control system is too "loose" to get any sort of useful trajectory correspondence, without restricting the class of vector fields giving rise to the differential inclusion. Roughly speaking, this is because a differential inclusion only prescribes the values of vector fields, and the topologies have to do with derivatives as well. ○

5.6 The Category of Tautological Control Systems

In our discussion of feedback equivalence in Sect. 1.1.2 we indicated that the notion of equivalence in our framework is not interesting to us. In this section, we illustrate why it not interesting by defining a natural notion of equivalence, and then seeing that it degenerates to something trivial under natural hypotheses. We do this in a general way by considering first how one might define a "category" of tautological control systems with objects and morphisms. The problem of equivalence is then the problem of understanding isomorphisms in this category. By imposing a naturality condition on morphisms via trajectories, we prove that isomorphisms are uniquely determined by diffeomorphisms of the underlying manifolds for the two tautological control systems. The notion of "direct image" we use here is common in sheaf theory, and we refer to, e.g., [16, Definition 2.3.1] for some discussion. However, by far the best presentation that we could find of direct images of presheaves such as we use here is in the online documentation [21, Tag 008C].

Let us first describe how to build maps between tautological control systems. This is done first by making the following definition.

Definition 5.34 (Direct image of tautological control systems). Let $m \in \mathbb{Z}_{\geq 0}$ and $m' \in \{0, \mathrm{lip}\}$, let $\nu \in \{m + m', \infty, \omega\}$, and let $r \in \{\infty, \omega\}$, as required. Let $\mathfrak{G} = (\mathsf{M}, \mathscr{F})$ be a C^ν-tautological control system, let N be C^r-manifold, and let

$\Phi \in \mathrm{C}^r(\mathsf{M};\mathsf{N})$. The ***direct image*** of $\mathfrak{G}$ by Φ is the tautological control system $\Phi_*\mathfrak{G} = (\mathsf{N}, \Phi_*\mathscr{F})$ defined by $\Phi_*\mathscr{F}(\mathcal{V}) = \mathscr{F}(\Phi^{-1}(\mathcal{V}))$ for $\mathcal{V} \subseteq \mathsf{N}$ open. ○

One easily verifies that if $\mathscr{F}$ is a sheaf, then so too is $\Phi_*\mathscr{F}$.

With the preceding sheaf construction, we can define what we mean by a morphism of tautological control systems.

Definition 5.35 (Morphism of tautological control systems). Let $m \in \mathbb{Z}_{\geq 0}$ and $m' \in \{0, \mathrm{lip}\}$, let $\nu \in \{m + m', \infty, \omega\}$, and let $r \in \{\infty, \omega\}$, as required. Let $\mathfrak{G} = (\mathsf{M}, \mathscr{F})$ and $\mathfrak{H} = (\mathsf{N}, \mathscr{G})$ be C^ν-tautological control systems. A ***morphism*** from $\mathfrak{G}$ to $\mathfrak{H}$ is a pair $(\Phi, \Phi^\sharp)$ such that

(i) $\Phi \in \mathrm{C}^r(\mathsf{M};\mathsf{N})$ and
(ii) $\Phi^\sharp = (\Phi^\sharp_{\mathcal{V}})_{\mathcal{V}\,\mathrm{open}}$ is a family of mappings $\Phi^\sharp_{\mathcal{V}} \colon \mathscr{G}(\mathcal{V}) \to \Phi_*\mathscr{F}(\mathcal{V})$, $\mathcal{V} \subseteq \mathsf{N}$ defined as follows:
   (a) there exists a family $L_{\mathcal{V}} \in \mathrm{L}(\Gamma^\nu(\mathsf{T}\mathcal{V}); \Gamma^\nu(\mathsf{T}(\Phi^{-1}(\mathcal{V}))))$ of continuous linear mappings satisfying $L_{\mathcal{V}'} = L_{\mathcal{V}}|\Gamma^\nu(\mathsf{T}\mathcal{V}')$ if $\mathcal{V}, \mathcal{V}' \subseteq \mathsf{N}$ are open with $\mathcal{V}' \subseteq \mathcal{V}$;
   (b) $\Phi^\sharp_{\mathcal{V}} = L_{\mathcal{V}}|\mathscr{G}(\mathcal{V})$. ○

By the preceding definition, we arrive at the "category of C^ν-tautological control systems" whose objects are tautological control systems and whose morphisms are as just defined. From the point of view of control theory, one wishes to restrict these definitions further to account for the fact that morphisms ought to preserve trajectories. Therefore, let us see how trajectories come into the picture. First we consider open-loop systems. Thus let $\mathbb{T}$ be a time-domain and let $\mathcal{V} \subseteq \mathsf{N}$ be open. If $Y \colon \mathbb{T} \to \mathscr{G}(\mathcal{U})$, then we have $\Phi^\sharp(Y)_t \triangleq \Phi^\sharp_{\mathcal{V}}(Y_t) \in \mathscr{F}(\Phi^{-1}(\mathcal{V}))$ for each $t \in \mathbb{T}$. That is, an open-loop system $(Y, \mathbb{T}, \mathcal{V})$ for $\mathfrak{H}$ gives rise to an open-loop system $(\Phi^\sharp(Y), \mathbb{T}, \Phi^{-1}(\mathcal{V}))$ for $\mathfrak{G}$. For such a correspondence to have significance, it must do the more or less obvious thing to trajectories.

Definition 5.36 (Trajectory-preserving morphisms of tautological control systems). Let $m \in \mathbb{Z}_{\geq 0}$ and $m' \in \{0, \mathrm{lip}\}$, let $\nu \in \{m + m', \infty, \omega\}$, and let $r \in \{\infty, \omega\}$, as required. Let $\mathfrak{G} = (\mathsf{M}, \mathscr{F})$ and $\mathfrak{H} = (\mathsf{N}, \mathscr{G})$ be C^ν-tautological control systems. A morphism $(\Phi, \Phi^\sharp)$ from $\mathfrak{G}$ to $\mathfrak{H}$ is ***trajectory-preserving*** if, for each time-domain $\mathbb{T}$, each open $\mathcal{V} \subseteq \mathsf{N}$, and each $Y \in \mathrm{LI}\Gamma^\nu(\mathbb{T}; \mathscr{G}(\mathcal{V}))$, any integral curve $\xi \colon \mathbb{T}' \to \Phi^{-1}(\mathcal{V})$ for the time-varying vector field $t \mapsto \Phi^\sharp(Y_t)$ defined on $\mathbb{T}' \subseteq \mathbb{T}$ has the property that $\Phi \circ \xi$ is an integral curve for Y. ○

Note that the time-varying vector field $t \mapsto \Phi^\sharp(Y_t)$ from the definition is locally integrally bounded by [2, Lemma 1.2].

We can now characterise these trajectory-preserving morphisms.

Proposition 5.37 (Characterisation of trajectory-preserving morphisms). *Let $m \in \mathbb{Z}_{\geq 0}$ and $m' \in \{0, \mathrm{lip}\}$, let $\nu \in \{m + m', \infty, \omega\}$, and let $r \in \{\infty, \omega\}$, as required. Let $\mathfrak{G} = (\mathsf{M}, \mathscr{F})$ and $\mathfrak{H} = (\mathsf{M}, \mathscr{G})$ be C^ν-tautological control systems. A morphism $(\Phi, \Phi^\sharp)$ from $\mathfrak{G}$ to $\mathfrak{H}$ is trajectory-preserving if and only if, for each open $\mathcal{V} \subseteq \mathsf{N}$, each $Y \in \mathscr{G}(\mathcal{V})$, each $y \in \mathcal{V}$, and each $x \in \Phi^{-1}(y)$, we have $T_x\Phi(\Phi^\sharp(Y)(x)) = Y(y)$.*

Proof First suppose that $(\Phi, \Phi^\sharp)$ is trajectory-preserving, and let $\mathcal{V} \subseteq \mathsf{N}$ be open, let $Y \in \mathscr{G}(\mathcal{V})$, let $y \in \mathcal{V}$, and let $x \in \Phi^{-1}(\mathcal{V})$. Let $\mathbb{T} \subseteq \mathbb{R}$ be a time-domain for which $0 \in \operatorname{int}(\mathbb{T})$ and for which the integral curve η for Y through y is defined on $\mathbb{T}$. We consider $Y \in \mathrm{LI}\Gamma^\nu(\mathbb{T}; \mathscr{G}(\mathcal{V}))$ by taking $Y_t = Y$, i.e., Y is a time-independent time-varying vector field. Note that integral curves of Y can, therefore, be chosen to be differentiable [6, Theorem 1.3], and *will be* differentiable if $\nu > 0$. Let $\mathbb{T}' \subseteq \mathbb{T}$ be such that the differentiable integral curve ξ for $\Phi^\sharp(Y)$ through x is defined on $\mathbb{T}'$. Since $(\Phi, \Phi^\sharp)$ is trajectory-preserving, we have $\eta = \Phi \circ \xi$ on $\mathbb{T}'$. Therefore,

$$Y(y) = \eta'(0) = T_x\Phi(\xi'(0)) = T_x\Phi(\Phi^\sharp(Y)(x)).$$

Next suppose that, for each open $\mathcal{V} \subseteq \mathsf{N}$, each $Y \in \mathscr{G}(\mathcal{V})$, each $y \in \mathcal{V}$, and each $x \in \Phi^{-1}(y)$, we have $T_x\Phi(\Phi^\sharp(Y)(x)) = Y(y)$. Let $\mathbb{T}$ be a time-domain, let $\mathcal{V} \subseteq \mathsf{N}$ be open, let $Y \in \mathrm{LI}\Gamma^\nu(\mathbb{T}; \mathscr{G}(\mathcal{V}))$, and let $\xi : \mathbb{T}' \to \Phi^{-1}(\mathcal{V})$ be an integral curve for the time-varying vector field $t \mapsto \Phi^\sharp(Y_t)$ defined on $\mathbb{T}' \subseteq \mathbb{T}$. Let $\eta = \Phi \circ \xi$. Then we have

$$\eta'(t) = T_{\xi(t)}\Phi(\Phi^\sharp(Y_t)(\xi(t))) = Y_t(\eta(t))$$

for almost every $t \in \mathbb{T}'$, showing that η is an integral curve for Y. □

Note that the condition $T_x\Phi(\Phi^\sharp(Y)(x)) = Y(y)$ is consistent with the regularity conditions for Y. In the cases $\nu \in \{m, \infty, \omega\}$, this is a consequence of the Chain Rule (see [17, Proposition 2.2.8] for the real analytic case). In the Lipschitz case this is a consequence of the fact that the Lipschitz constant of a differentiable map is the norm of the derivative [10, Example 1.4(c)], combined with the fact that the Lipschitz constant of a composition is the product of the Lipschitz constants [22, Proposition 1.2.2].

To make a connection with more common notions of mappings between control systems, let us do the following. Let $m \in \mathbb{Z}_{\geq 0}$ and $m' \in \{0, \mathrm{lip}\}$, let $\nu \in \{m + m', \infty, \omega\}$, and let $r \in \{\infty, \omega\}$, as required. Suppose that we have two C^ν-control systems $\Sigma_1 = (\mathsf{M}_1, F_1, \mathcal{C}_1)$ and $\Sigma_2 = (\mathsf{M}_2, F_2, \mathcal{C}_2)$. As tautological control systems, these are globally generated, so let us not fuss with general open sets for the purpose of this illustrative discussion. We then suppose that we have a mapping $\Phi \in \mathrm{C}^r(\mathsf{M}_1; \mathsf{M}_2)$ and a mapping $\kappa : \mathsf{M}_1 \times \mathcal{C}_2 \to \mathcal{C}_1$, which gives rise to a correspondence between the system vector fields by

$$\Phi^\sharp(F_2^{u_2})(x_1) = F_1^{\kappa(x_1, u_2)}(x_1).$$

The condition of being trajectory-preserving means that a trajectory ξ_1 for Σ_1 satisfying

$$\xi_1'(t) = F_1(\xi_1(t), \kappa(\xi_1(t), \mu_2(t)))$$

gives rise to a trajectory $\xi_2 = \Phi \circ \xi_1$ for Σ_2, implying that

$$\xi_2' = T_{\xi_1(t)}\Phi(\xi_1'(t)) = T_{\xi_1(t)}\Phi \circ F_1(\xi_1(t), \kappa(\xi_1(t), \mu_2(t))).$$

Thus

$$F_2(x_2, u_2) = T_{x_1}\Phi \circ F_1(x_1, \kappa(x_1, u_2))$$

for every $x_1 \in \Phi^{-1}(x_2)$.

There may well be some interest in studying general morphisms, but we will not pursue this right at the moment. Instead, let us simply think about isomorphisms in the category of tautological control systems.

Definition 5.38 (Isomorphisms of tautological control systems). Let $m \in \mathbb{Z}_{\geq 0}$ and $m' \in \{0, \text{lip}\}$, let $\nu \in \{m + m', \infty, \omega\}$, and let $r \in \{\infty, \omega\}$, as required. Let $\mathfrak{G} = (\mathsf{M}, \mathscr{F})$ and $\mathfrak{H} = (\mathsf{N}, \mathscr{G})$ be C^ν-tautological control systems. An ***isomorphism*** from $\mathfrak{G}$ to $\mathfrak{H}$ is a morphism $(\Phi, \Phi^\sharp)$ such that Φ is a diffeomorphism and $L_{\mathcal{V}}$ is an isomorphism (in the category of locally convex topological vector spaces) for every open $\mathcal{V} \subseteq \mathsf{N}$, where $L_{\mathcal{V}}$ is such that $\Phi^\sharp_{\mathcal{V}} = L_{\mathcal{V}}|\mathscr{G}(\mathcal{V})$ as in Definition 5.35. ◦

It is now easy to describe the trajectory-preserving isomorphisms.

Proposition 5.39 (Characterisation of trajectory-preserving isomorphisms). *Let $m \in \mathbb{Z}_{\geq 0}$ and $m' \in \{0, \text{lip}\}$, let $\nu \in \{m+m', \infty, \omega\}$, and let $r \in \{\infty, \omega\}$, as required. Let $\mathfrak{G} = (\mathsf{M}, \mathscr{F})$ and $\mathfrak{H} = (\mathsf{N}, \mathscr{G})$ be C^ν-tautological control systems. A morphism $(\Phi, \Phi^\sharp)$ from $\mathfrak{G}$ to $\mathfrak{H}$ is a trajectory-preserving isomorphism if and only if Φ is a diffeomorphism and*

$$\mathscr{G}(\Phi(\mathcal{U})) = \{(\Phi|\mathcal{U})_* X \mid X \in \mathscr{F}(\mathcal{U})\}$$

for every open set $\mathcal{U} \subseteq \mathsf{M}$.

Proof According to Proposition 5.37, if $\mathcal{V} \subseteq \mathsf{N}$ is open and if $Y \in \mathscr{G}(\mathcal{V})$, we have $(\Phi|\Phi^{-1}(\mathcal{V}))_*(\Phi^\sharp(Y)) = Y$ or $\Phi^\sharp(Y) = (\Phi|\Phi^{-1}(\mathcal{V}))^* Y$. Since $\Phi^\sharp$ is a bijection from $\mathscr{G}(\mathcal{V})$ to $\mathscr{F}(\Phi^{-1}(\mathcal{V}))$, we conclude that

$$\mathscr{F}(\Phi^{-1}(\mathcal{V})) = \{(\Phi|\Phi^{-1}(\mathcal{V}))^* Y \mid Y \in \mathscr{G}(\mathcal{V})\}.$$

This is clearly equivalent to the assertion of the theorem since Φ must be a diffeomorphism. □

In words, trajectory-preserving isomorphisms simply amount to the natural correspondence of vector fields under the push-forward Φ_*. (One should verify that push-forward is continuous as a mapping between locally convex spaces. This amounts to proving continuity of composition, and for this we point to places in the literature from which this can be deduced. In the smooth and finitely differentiable cases this can be shown using an argument fashioned after that from [18, Proposition 1]. In the Lipschitz case, this follows because the Lipschitz constant of a composition is bounded by the product of the Lipschitz constants [22, Proposition 1.2.2]. In the real analytic case, this follows from Sublemma 6 from the proof of [15, Lemma 2.4].) In particular, if one wishes to consider only the identity diffeomorphism, i.e., only consider the "feedback part" of a feedback transformation, we see that the only trajectory-preserving isomorphism is simply the identity morphism. In this way we see that the notion of equivalence for tautological control systems is either very trivial (it is easy to understand when systems are equivalent) or very difficult (the study of equivalence classes contains as a special case the classification of vector fields up to diffeomorphism), depending on your tastes. It is our view that the triviality (or impossibility) of equivalence is a virtue of the formulation since all structure except that of the manifold and the vector fields has been removed; there is no extraneous structure. This justifies our calling these "tautological" control systems. We refer to Sect. 1.1.2 for further discussion.

References

1. Abraham R, Marsden JE, Ratiu TS (1988) Manifolds, tensor analysis, and applications, 2 edn. No. 75 in Applied Mathematical Sciences. Springer, Berlin
2. Beckmann R, Deitmar A (2011) Strong vector valued integrals. ArXiv:1102.1246v1 [math.FA]. http://arxiv.org/abs/1102.1246v1
3. Bogachev VI (2007) Measure theory, vol 2. Springer, New York
4. Bourbaki N (1989) General topology I. Elements of Mathematics. Springer, New York
5. Cartan H (1957) Variétés analytiques réelles et variétés analytiques complexes. Bull. Soc. Math. France 85:77–99
6. Coddington EE, Levinson N (1984) Theory of ordinary differential equations, 8th edn. Robert E. Krieger Publishing Company, Huntington
7. Cohn DL (1980) Measure theory. Birkhäuser, Boston
8. Demailly JP (2012) Complex analytic and differential geometry. Unpublished manuscript made publicly available. http://www-fourier.ujf-grenoble.fr/~demailly/manuscripts/agbook.pdf
9. Elliott DL (2009) Bilinear systems. Matrices in action. No. 169 in Applied Mathematical Sciences. Springer, New York
10. Gromov M (1999) Metric structures for Riemannian and non-Riemannian spaces. Modern Birkhäuser Classics. Birkhäuser, Boston
11. Gunning RC (1990) Introduction to holomorphic functions of several variables, vol I: function theory. Wadsworth & Brooks/Cole Mathematics Series. Wadsworth & Brooks/Cole, Belmont
12. Himmelberg CJ (1975) Measurable relations. Fund Math 87:53–72
13. Horváth J (1966) Topological vector spaces and distributions, vol I. Addison Wesley, Reading
14. Jafarpour S, Lewis AD (2014) Locally convex topologies and control theory. Submitted to SIAM J Control Optim
15. Jafarpour S, Lewis AD (2014) Time-varying vector fields and their flows. To appear in Springer Briefs in Mathematics
16. Kashiwara M, Schapira P (1990) Sheaves on manifolds. No. 292 in Grundlehren der Mathematischen Wissenschaften. Springer, New York
17. Krantz SG, Parks HR (2002) A primer of real analytic functions, 2nd edn. Birkhäuser Advanced Texts. Birkhäuser, Boston
18. Mather JN (1969) Stability of C^{∞}-mappings: II. Infinitesimal stability implies stability. Ann Math 89(2):254–291
19. Pietsch A (1969) Nuclear locally convex spaces. No. 66 in Ergebnisse der Mathematik und ihrer Grenzgebiete. Springer, New York

20. Schaefer HH, Wolff MP (1999) Topological vector spaces, 2 edn. No. 3 in Graduate Texts in Mathematics. Springer, New York
21. Stacks Project Authors (2014) Stacks Project. http://stacks.math.columbia.edu
22. Weaver N (1999) Lipschitz algebras. World Scientific, Singapore
23. Willard S (2004) General topology. Dover Publications Inc, New York. Reprint of 1970 Addison-Wesley edition

Chapter 6
Étalé Systems

The development of tautological control systems in the preceding chapter was focussed in large part on connecting this new class of control systems with more common existing classes of systems. In particular, our notion of a trajectory is a quite natural adaptation to our framework of the usual notion of a trajectory for a control system. However, it turns out that there is a limitation of this sort of definition in terms of being able to use the full power of the tautological control system framework. In this chapter we overcome this limitation, and at the same time more fully integrate the sheaf formalism into the way in which we think about tautological control systems.

In order to begin to understand this, let us consider an example.

Example 6.1 (A different sort of trajectory). We take $\mathsf{M} = \mathbb{R}^2$ with coordinates (x, y) and let X and Y be the vector fields

$$X = \frac{\partial}{\partial x}, \quad Y = \frac{\partial}{\partial y}.$$

We let $\mathfrak{G} = (\mathsf{M}, \mathscr{F})$ be the tautological control system defined by taking

$$\mathscr{F}(\mathcal{U}) = \begin{cases} \{\pm X|\mathcal{U}, \pm Y|\mathcal{U}\}, & \mathcal{U} \subseteq (-1, 1) \times \mathbb{R}, \\ \{\pm X|\mathcal{U}\}, & \text{otherwise}. \end{cases}$$

It is a simple verification to see that $\mathscr{F}$ is a presheaf of sets of C^ω-vector fields. Consider the curve $\xi : [0, 6] \to \mathsf{M}$ defined by

A.D. Lewis, *Tautological Control Systems*, SpringerBriefs in Control, Automation and Robotics, DOI: 10.1007/978-3-319-08638-5_6

$$\xi(t) = \begin{cases} (2-t,-1), & t \in [0,2], \\ (0,t-3), & t \in (2,4], \\ (t-4,1), & t \in (4,6]. \end{cases}$$

Note that ξ cannot be a trajectory for the tautological control system $\mathfrak{G}$ according to Definition 5.14 because any open set $\mathcal{U} \subseteq \mathsf{M}$ containing image(ξ) will have the property that $Y|\mathcal{U} \notin \mathscr{F}(\mathcal{U})$. ∘

It is pretty clear that, despite the fact that the curve of the example is not a trajectory according to our existing definition, we would like it to be a trajectory. The idea of a more general notion of a trajectory is that it should *locally* be an integral curve for some system vector field. The notion of "local" for our system is captured by stalks, and this leads us to the consideration of the étalé space $\mathrm{Et}(\mathscr{F})$ as a device for capturing trajectories of the sort depicted in Example 6.1. There are some technicalities that have to be dealt with to provide a satisfactory description of what is required, and in this chapter we undertake this development.

6.1 Sheaves of Time-Varying Vector Fields

In the next definition we introduce some sheaves of time-varying vector fields. These are not defined by simply defining the presheaf, but by a more indirect construction; they are constructed by defining their local sections over a basis for the topology consisting of products of open subsets of $\mathbb{T}$ and M. The idea is that, to every open set $\mathcal{W} \subseteq \mathbb{T} \times \mathsf{M}$, we assign a "time-varying vector field". However, this vector field is not defined on a single time interval, but rather it is defined locally in M on a variable interval. That this procedure produces a legitimate sheaf is proved in [3, Theorem II.1.3] (see also the detailed discussion in the open source book [5, Tag 009H]).

With the preceding as preparation, we state the following definition.

Definition 6.2 (Sheaves of time-varying vector fields). Let $m \in \mathbb{Z}_{\geq 0}$ and $m' \in \{0, \mathrm{lip}\}$, let $\nu \in \{m+m', \infty, \omega\}$, and let $r \in \{\infty, \omega\}$, as required. Let $\mathbb{T} \subseteq \mathbb{R}$ be an interval and let M be a C^r-manifold.

(i) By $\mathrm{CF}\mathscr{G}^\nu(\mathbb{T}; \mathsf{TM})$ we denote the sheaf over $\mathbb{T} \times \mathsf{M}$ defined by requiring that

$$\mathrm{CF}\mathscr{G}^\nu(\mathbb{T}; \mathsf{TM})(\mathbb{T}' \times \mathcal{U}) = \mathrm{CF}(\mathbb{T}'; \Gamma^\nu(\mathsf{T}\mathcal{U}))$$

for every relatively open interval $\mathbb{T}' \subseteq \mathbb{T}$ and every open set $\mathcal{U} \subseteq \mathsf{M}$. Here $\mathrm{CF}(\mathbb{T}'; \Gamma^\nu(\mathsf{T}\mathcal{U}))$ denotes the set of measurable functions from $\mathbb{T}'$ to $\Gamma^\nu(\mathsf{T}\mathcal{U})$.

(ii) By $\mathrm{LI}\mathscr{G}^\nu(\mathbb{T}; \mathsf{TM})$ we denote the sheaf over $\mathbb{T} \times \mathsf{M}$ defined by requiring that

$$\mathrm{LI}\mathscr{G}^\nu(\mathbb{T}; \mathsf{TM})(\mathbb{T}' \times \mathcal{U}) = \mathrm{L}^1(\mathbb{T}'; \Gamma^\nu(\mathsf{T}\mathcal{U}))$$

for every relatively open interval $\mathbb{T}' \subseteq \mathbb{T}$ and every open set $\mathcal{U} \subseteq \mathsf{M}$. Here $\mathrm{L}^1(\mathbb{T}'; \Gamma^\nu(\mathsf{T}\mathcal{U}))$ denotes the set of Bochner integrable functions from $\mathbb{T}'$ to $\Gamma^\nu(\mathsf{T}\mathcal{U})$.

(iii) By $\mathrm{LB}\mathscr{G}^\nu(\mathbb{T}; \mathsf{TM})$ we denote the sheaf over $\mathbb{T} \times \mathsf{M}$ defined by requiring that

$$\mathrm{LB}\mathscr{G}^\nu(\mathbb{T}; \mathsf{TM})(\mathbb{T}' \times \mathcal{U}) = \mathrm{L}^\infty(\mathbb{T}'; \Gamma^\nu(\mathsf{T}\mathcal{U}))$$

for every relatively open interval $\mathbb{T}' \subseteq \mathbb{T}$ and every open set $\mathcal{U} \subseteq \mathsf{M}$. Here $\mathrm{L}^\infty(\mathbb{T}'; \Gamma^\nu(\mathsf{T}\mathcal{U}))$ denotes the set of essentially von Neumann bounded functions from $\mathbb{T}'$ to $\Gamma^\nu(\mathsf{T}\mathcal{U})$. ∘

We remind the reader of the characterisations of $\mathrm{CF}(\mathbb{T}'; \Gamma^\nu(\mathsf{T}\mathcal{U}))$, $\mathrm{L}^1(\mathbb{T}'; \Gamma^\nu(\mathsf{T}\mathcal{U}))$, and $\mathrm{L}^\infty(\mathbb{T}'; \Gamma^\nu(\mathsf{T}\mathcal{U}))$ made possible by Theorem 3.11.

In terms of sheaf theory, the sheaf $\mathrm{CF}(\mathbb{T}'; \Gamma^\nu(\mathsf{T}\mathcal{U}))$ is to be thought of as a sheaf in the category of $\mathbb{R}$-vector spaces, and $\mathrm{L}^1(\mathbb{T}'; \Gamma^\nu(\mathsf{T}\mathcal{U}))$ and $\mathrm{L}^\infty(\mathbb{T}'; \Gamma^\nu(\mathsf{T}\mathcal{U}))$ are to be thought of as sheaves in the category of locally convex topological vector spaces. That is, to each open $\mathcal{W} \subseteq \mathbb{T} \times \mathsf{M}$ we assign a (locally convex topological) vector space.

The main point that we shall use is that, as sheaves, the local sections of $\mathrm{CF}\mathscr{G}^\nu(\mathbb{T}; \mathsf{TM})$ (resp. $\mathrm{LI}\mathscr{G}^\nu(\mathbb{T}; \mathsf{TM})$, $\mathrm{LB}\mathscr{G}^\nu(\mathbb{T}; \mathsf{TM})$) are to be regarded as local sections of the corresponding étalé spaces, cf. Remark 4.10–3. Thus a local section $\mathscr{X}$ over an open set $\mathcal{W} \subseteq \mathbb{T} \times \mathsf{M}$ should be thought of as assigning to $(t, x) \in \mathcal{W}$ a germ $\mathscr{X}(t, x) \in \mathrm{Et}(\mathrm{CF}\mathscr{G}^\nu(\mathbb{T}; \mathsf{TM}))_{(t,x)}$ (resp. $\mathrm{Et}(\mathrm{LI}\mathscr{G}^\nu(\mathbb{T}; \mathsf{TM}))_{(t,x)}$, $\mathrm{Et}(\mathrm{LB}\mathscr{G}^\nu(\mathbb{T}; \mathsf{TM}))_{(t,x)}$). The definitions of these sheaves permit the following characterisation of germs.

Construction 6.3 (Representatives of germs for sheaves of time-varying vector fields). Because of the way the sheaves $\mathrm{CF}\mathscr{G}^\nu(\mathbb{T}; \mathsf{TM})$, $\mathrm{LI}\mathscr{G}^\nu(\mathbb{T}; \mathsf{TM})$, and $\mathrm{LB}\mathscr{G}^\nu(\mathbb{T}; \mathsf{TM})$ are defined by defining them on a basis for the topology of $\mathbb{T} \times \mathsf{M}$, for each germ $\mathscr{X}(t, x)$ there is a relatively open interval $\mathbb{T}' \subseteq \mathbb{T}$, an open set $\mathcal{U} \subseteq \mathsf{M}$, and $X \in \mathrm{CF}(\mathbb{T}'; \Gamma^\nu(\mathsf{T}\mathcal{U}))$ (resp. $X \in \mathrm{L}^1(\mathbb{T}'; \Gamma^\nu(\mathsf{T}\mathcal{U}))$, $X \in \mathrm{L}^\infty(\mathbb{T}'; \Gamma^\nu(\mathsf{T}\mathcal{U}))$) such that $\mathbb{T}' \times \mathcal{U} \subseteq \mathcal{W}$ is a neighbourhood of (t, x) and such that $\mathscr{X}(t, x) = [X]_{(t,x)}$. ∘

As sheaves, $\mathrm{CF}\mathscr{G}^\nu(\mathbb{T}; \mathsf{TM})$, $\mathrm{LI}\mathscr{G}^\nu(\mathbb{T}; \mathsf{TM})$, and $\mathrm{LB}\mathscr{G}^\nu(\mathbb{T}; \mathsf{TM})$ are equipped with restriction maps

$$r_{\mathcal{W}_2,\mathcal{W}_1} : \mathrm{CF}\mathscr{G}^\nu(\mathbb{T}; \mathsf{TM})(\mathcal{W}_2) \to \mathrm{CF}\mathscr{G}^\nu(\mathbb{T}; \mathsf{TM})(\mathcal{W}_1),$$

and similarly for $\mathrm{LI}\mathscr{G}^\nu(\mathbb{T}; \mathsf{TM})$ and $\mathrm{LB}\mathscr{G}^\nu(\mathbb{T}; \mathsf{TM})$. To make these restriction maps explicit, we note that, since $\mathrm{CF}\mathscr{G}^\nu(\mathbb{T}; \mathsf{TM})$, $\mathrm{LI}\mathscr{G}^\nu(\mathbb{T}; \mathsf{TM})$, and $\mathrm{LB}\mathscr{G}^\nu(\mathbb{T}; \mathsf{TM})$ are sheaves, local sections are identified with local sections of the étalé space. In this case, the restriction maps are just regular restriction.

Although a local section of $\mathrm{CF}\mathscr{G}^\nu(\mathbb{T}; \mathsf{TM})$, $\mathrm{LI}\mathscr{G}^\nu(\mathbb{T}; \mathsf{TM})$, or $\mathrm{LB}\mathscr{G}^\nu(\mathbb{T}; \mathsf{TM})$ is not a time-varying vector field in the usual sense, we can nonetheless assign tangent vectors as if it is in the following manner (we do this for $\mathrm{CF}\mathscr{G}^\nu(\mathbb{T}; \mathsf{TM})$, but the same constructions apply to $\mathrm{LI}\mathscr{G}^\nu(\mathbb{T}; \mathsf{TM})$ and $\mathrm{LB}\mathscr{G}^\nu(\mathbb{T}; \mathsf{TM})$). Let $\mathcal{W} \subseteq \mathbb{T} \times \mathsf{M}$ be open and let $(t, x) \in \mathcal{W}$. If $\mathscr{X} \in \mathrm{CF}\mathscr{G}^\nu(\mathbb{T}; \mathsf{TM})(\mathcal{W})$ then, as in Construction 6.3, $\mathscr{X}$ is the germ of a time-varying vector field $X \in \mathrm{CF}(\mathbb{T}'; \Gamma^\nu(\mathsf{T}\mathcal{U}))$ for some relatively open interval $\mathbb{T}' \subseteq \mathbb{T}$ and some open set $\mathcal{U} \subseteq \mathsf{M}$ for which $\mathbb{T}' \times \mathcal{U} \subseteq \mathcal{W}$ is a neighbourhood of (t, x). With this in mind, we define

$$\begin{aligned} \mathrm{ev}_{(t,x)} \colon \mathrm{Et}(\mathrm{CF}\mathscr{G}^\nu(\mathbb{T}; \mathsf{TM}))_{(t,x)} &\to \mathsf{T}_x\mathsf{M} \\ \mathscr{X}(t, x) &\mapsto X(t, x). \end{aligned}$$

6.2 An Alternative Description of Local Sections of Sheaves of Time-Varying Vector Fields

The preceding constructions are a little bit hampered by the fact that the sheaves $\mathrm{CF}\mathscr{G}^\nu(\mathbb{T}; \mathsf{TM})$, $\mathrm{LI}\mathscr{G}^\nu(\mathbb{T}; \mathsf{TM})$, and $\mathrm{LB}\mathscr{G}^\nu(\mathbb{T}; \mathsf{TM})$ are not really comprised of time-varying vector fields, and so representations of their local sections can be a little awkward. Let us address this by showing that one can equivalently characterise sections of these sheaves as mappings into $\mathrm{Et}(\mathscr{G}^\nu_{\mathsf{TM}})$.

The ideas here mirror, to some extent, the constructions of Sect. 3.1. To this end, we begin with the following constructions that are to be regarded as the étalé versions of the corresponding notions for time-varying vector fields. In the following definition, we suppose that $\mathrm{Et}(\mathscr{G}^\nu_{\mathsf{TM}})$ is equipped with the étalé topology (from Definition 4.9) and that the stalks $\mathscr{G}^\nu_{x,\mathsf{TM}}$ are equipped with the C^ν-stalk topology (from Definition 4.11).

Definition 6.4 (Time-varying local sections of $\mathscr{G}^\nu_{\mathsf{TM}}$). Let $m \in \mathbb{Z}_{\geq 0}$ and $m' \in \{0, \mathrm{lip}\}$, let $\nu \in \{m + m', \infty, \omega\}$, and let $r \in \{\infty, \omega\}$, as required. Let $\mathbb{T} \subseteq \mathbb{R}$ be an interval, let M be a C^r-manifold, let $\mathcal{W} \subseteq \mathbb{T} \times \mathsf{M}$ be open, and let $\mathscr{X} \colon \mathcal{W} \to \mathrm{Et}(\mathscr{G}^\nu_{\mathsf{TM}})$ be such that $\mathscr{X}(t, x) \in \mathscr{G}^\nu_{x,\mathsf{TM}}$. For $(t, x) \in \mathcal{W}$, denote

$$\mathcal{W}_t = \{x \in \mathsf{M} \mid (t, x) \in \mathcal{W}\}, \quad \mathcal{W}^x = \{t \in \mathbb{T} \mid (t, x) \in \mathcal{W}\},$$

and define mappings

$$\begin{aligned} \mathscr{X}_t \colon \mathcal{W}_t &\to \mathrm{Et}(\mathscr{G}^\nu_{\mathsf{TM}}) & \mathscr{X}^x \colon \mathcal{W}^x &\to \mathscr{G}^\nu_{x,\mathsf{TM}} \\ x &\mapsto \mathscr{X}(t, x), & t &\mapsto \mathscr{X}(t, x). \end{aligned}$$

(i) The mapping $\mathscr{X}$ is a ***Carathéodory local section of class*** $\mathbf{C}^\nu$ if

  (a) for each $t \in \mathbb{T}$ for which $\mathcal{W}_t \neq \emptyset$, $\mathscr{X}_t$ is continuous and
  (b) for each $x \in \mathsf{M}$ for which $\mathcal{W}^x \neq \emptyset$, $\mathscr{X}^x$ is measurable.

(ii) The mapping $\mathscr{X}$ is a ***locally integrally bounded local section of class*** $\mathbf{C}^\nu$ if

  (a) for each $t \in \mathbb{T}$ for which $\mathcal{W}_t \neq \emptyset$, $\mathscr{X}_t$ is continuous and
  (b) for each $x \in \mathsf{M}$ for which $\mathcal{W}^x \neq \emptyset$, $\mathscr{X}^x$ is locally Bochner integrable.

(iii) The mapping $\mathscr{X}$ is a ***locally essentially von Neumann bounded local section of class*** C^ν if

(a) for each $t \in \mathbb{T}$ for which $\mathcal{W}_t \neq \emptyset$, $\mathscr{X}_t$ is continuous and
(b) for each $x \in \mathsf{M}$ for which $\mathcal{W}^x \neq \emptyset$, $\mathscr{X}^x$ is locally essentially von Neumann bounded.

We wish to establish a correspondence between the objects in the preceding definition and those in Definition 6.2. To do this we shall need some general notation to capture a notion where both sorts of objects have no a priori structure in terms of their time-dependence. To this end, let $m \in \mathbb{Z}_{\geq 0}$ and $m' \in \{0, \mathrm{lip}\}$, let $\nu \in \{m+m', \infty, \omega\}$, and let $r \in \{\infty, \omega\}$, as required. Let $\mathbb{T} \subseteq \mathbb{R}$ be an interval and let M be a C^r-manifold. We let $\mathrm{TV}\mathscr{G}^\nu(\mathbb{T}; \mathsf{TM})$ be the sheaf over $\mathbb{T} \times \mathsf{M}$ defined by requiring that

$$\mathrm{TV}\mathscr{G}^\nu(\mathbb{T}; \mathsf{TM})(\mathbb{T}' \times \mathcal{U}) = \mathrm{Map}(\mathbb{T}'; \Gamma^\nu(\mathsf{T}\mathcal{U})),$$

for a relatively open interval $\mathbb{T}' \subseteq \mathbb{T}$ and an open set $\mathcal{U} \subseteq \mathsf{M}$, and where Map simply means the set of all maps. This is a sheaf in the same manner as are the sheaves $\mathrm{CF}\mathscr{G}^\nu(\mathbb{T}; \mathsf{TM})$, $\mathrm{LI}\mathscr{G}^\nu(\mathbb{T}; \mathsf{TM})$, and $\mathrm{LB}\mathscr{G}^\nu(\mathbb{T}; \mathsf{TM})$, being defined on a basis for the topology of $\mathbb{T} \times \mathsf{M}$. It is, moreover, a sheaf in the category of $\mathbb{R}$-vector spaces. In particular, Construction 6.3 applies to describe a representative of a germ for this sheaf. For $\mathcal{W} \subseteq \mathbb{T} \times \mathsf{M}$ open, we also denote by $\mathrm{Map}_0(\mathcal{W}; \mathrm{Et}(\mathscr{G}^\nu_{\mathsf{TM}}))$ the mappings $\mathscr{X} : \mathcal{W} \to \mathrm{Et}(\mathscr{G}^\nu_{\mathsf{TM}})$ such that $\mathscr{X}(t, x) \in \mathscr{G}^\nu_{x,\mathsf{TM}}$ for every $(t, x) \in \mathcal{W}$. We shall make use of the evaluation map

$$\begin{aligned} \mathrm{ev}_x : \mathscr{G}^\nu_{x,\mathsf{TM}} &\to \mathsf{T}_x\mathsf{M} \\ [X]_x &\mapsto X(x). \end{aligned}$$

We now indicate how to relate sections of $\mathrm{TV}\mathscr{G}^\nu(\mathbb{T}; \mathsf{TM})$ over $\mathcal{W}$ to elements of $\mathrm{Map}_0(\mathcal{W}; \mathrm{Et}(\mathscr{G}^\nu_{\mathsf{TM}}))$, and vice versa.

First let $\mathcal{W} \subseteq \mathbb{T} \times \mathsf{M}$ be open and let $\mathscr{X} \in \mathrm{TV}\mathscr{G}^\nu(\mathbb{T}; \mathsf{TM})(\mathcal{W})$. We define $\check{\mathscr{X}} \in \mathrm{Map}_0(\mathcal{W}; \mathrm{Et}(\mathscr{G}^\nu_{\mathsf{TM}}))$ as follows. Let $(t, x) \in \mathcal{W}$ and let $\mathbb{T}'$, $\mathcal{U}$, and $X \in \mathrm{Map}(\mathbb{T}'; \Gamma^\nu(\mathsf{T}\mathcal{U}))$ be as in Construction 6.3, i.e., such that $\mathscr{X}(t, x) = [X]_{(t,x)}$. Then define $\check{\mathscr{X}}(t, x) = [X_t]_x$, where $X_t(x') = X(t, x')$ for $x' \in \mathcal{U}$. This construction is well-defined in that (1) $X_t \in \Gamma^\nu(\mathsf{T}\mathcal{U})$ so the germ $[X_t]_x$ is in $\mathscr{G}^\nu_{x,\mathsf{TM}}$ and (2) it is independent of the choice of representative X.

Next let $\mathcal{W} \subseteq \mathbb{T} \times \mathsf{M}$ be open and let $\mathscr{X} \in \mathrm{Map}_0(\mathcal{W}; \mathrm{Et}(\mathscr{G}^\nu_{\mathsf{TM}}))$. We define $\hat{\mathscr{X}} \in \mathrm{TV}\mathscr{G}^\nu(\mathbb{T}; \mathsf{TM})(\mathcal{W})$ as follows. Let $(t, x) \in \mathcal{W}$, let $\mathbb{T}' \subseteq \mathbb{T}$ be a relatively open interval, and let $\mathcal{U} \subseteq \mathsf{M}$ be an open set such that $\mathbb{T}' \times \mathcal{U} \subseteq \mathcal{W}$ is a neighbourhood of (t, x). Denote

$$\begin{aligned} X : \mathbb{T}' \times \mathcal{U} &\to \mathsf{T}\mathcal{U} \\ (t', x') &\mapsto \mathrm{ev}_{x'}(\mathscr{X}(t', x')). \end{aligned}$$

It is clear that $X \in \operatorname{Map}(\mathbb{T}'; \Gamma^{\nu}(\mathsf{T}\mathcal{U}))$. We thus can define $\hat{\mathscr{X}}(t,x) = [X]_{(t,x)}$. We note that $\hat{\mathscr{X}}$ is well-defined in that (1) $[X]_{(t,x)} \in \operatorname{Et}(\mathsf{TV}\mathscr{G}^{\nu}(\mathbb{T}; \mathsf{TM}))_{(t,x)}$ and (2) the definition is independent of the representative X.

Note that the preceding correspondences are, in essence, making the identification of $[X_t]_x$ with $[X]_{(t,x)}$ for a suitable representative X of the germ.

With this notation, we can state and prove the following result.

Theorem 6.5 (A characterisation of local sections). *Let $m \in \mathbb{Z}_{\geq 0}$ and $m' \in \{0, \mathrm{lip}\}$, let $\nu \in \{m + m', \infty, \omega\}$, and let $r \in \{\infty, \omega\}$, as required. Let $\mathbb{T} \subseteq \mathbb{R}$ be an interval, let M be a C^r-manifold, let $\mathcal{W} \subseteq \mathbb{T} \times \mathsf{M}$ be open, and let $\mathscr{X} \in \operatorname{Map}_0(\mathcal{W}; \operatorname{Et}(\mathscr{G}^{\nu}_{\mathsf{TM}}))$. Then $\mathscr{X}$ is:*

(i) a Carathéodory local section of class C^{ν} if and only if $\hat{\mathscr{X}} \in \mathrm{CF}\mathscr{G}^{\nu}(\mathbb{T}; \mathsf{TM})(\mathcal{W})$;

(ii) a locally integrally bounded local section of class C^{ν} if and only if $\hat{\mathscr{X}} \in \mathrm{LI}\mathscr{G}^{\nu}(\mathbb{T}; \mathsf{TM})(\mathcal{W})$;

(iii) a locally essentially von Neumann bounded local section of class C^{ν} if and only if $\hat{\mathscr{X}} \in \mathrm{LB}\mathscr{G}^{\nu}(\mathbb{T}; \mathsf{TM})(\mathcal{W})$.

Proof (i) Suppose that $\hat{\mathscr{X}} \in \mathrm{CF}\mathscr{G}^{\nu}(\mathbb{T}; \mathsf{TM})(\mathcal{W})$. Let $(t,x) \in \mathcal{W}$, and let $\mathbb{T}'$, $\mathcal{U}$, and X be as in Construction 6.3. By definition of $\hat{\mathscr{X}}$, this implies that $X \in \operatorname{Map}(\mathbb{T}'; \Gamma^{\nu}(\mathsf{T}\mathcal{U}))$. As we asserted in Remark 4.10–3, the mapping $x' \mapsto \mathscr{X}(t,x') = [X_t]_x$ is continuous at x. Also, by definition of the stalk topology, the mapping

$$\begin{aligned} r_{\mathcal{U},x} \colon \mathscr{G}^{\nu}_{\mathsf{TM}}(\mathcal{U}) &\to \mathscr{G}^{\nu}_{x,\mathsf{TM}} \\ Y &\mapsto [Y]_x \end{aligned}$$

is continuous. Therefore, since $\mathscr{X}^x(t') = r_{\mathcal{U},x} \circ X_{t'}$ for $t' \in \mathcal{W}^x$ and since $t' \mapsto X_{t'}$ is measurable by assumption, it follows that $\mathscr{X}$ is a Carathéodory local section of class C^r.

Next suppose that $\mathscr{X}$ is a Carathéodory local section of class C^{ν}. Let $\mathbb{T}' \subseteq \mathbb{T}$ be a relatively open interval and let $\mathcal{U} \subseteq \mathsf{M}$ be open and such that $\mathbb{T}' \times \mathcal{U} \subseteq \mathcal{W}$, and let $X \in \operatorname{Map}(\mathbb{T}'; \Gamma^{\nu}(\mathsf{T}\mathcal{U}))$ be such that $\mathscr{X}(t,x) = [X_t]_x$. Since $\mathscr{X}$ is a Carathéodory local section, the map $t' \mapsto \mathscr{X}(t',x)$ is measurable. Since $\mathrm{ev}_x \colon \mathscr{G}^{\nu}_{x,\mathsf{TM}} \to \mathsf{T}_x\mathsf{M}$ is continuous (cf. the proof of Theorem 6.3 in [2]), it follows that the mapping

$$\mathbb{T}' \ni t' \mapsto \mathrm{ev}_x \circ [X_{t'}]_x = X^x(t') \in \mathsf{T}_x\mathsf{M}$$

is measurable. By Theorem 3.11 we conclude that $X \in \mathrm{CF}(\mathbb{T}'; \Gamma^{\nu}(\mathsf{TM}))$, implying that $\hat{\mathscr{X}} \in \mathrm{CF}\mathscr{G}^{\nu}(\mathbb{T}; \mathsf{TM})$.

(ii) Suppose that $\hat{\mathscr{X}} \in \mathrm{LI}\mathscr{G}^{\nu}(\mathbb{T}; \mathsf{TM})(\mathcal{W})$. Let $(t,x) \in \mathcal{W}$, and let $\mathbb{T}'$, $\mathcal{U}$, and X be as in Construction 6.3. As in the proof of part (i), we can conclude that $\mathscr{X}_t$ is continuous at x. Since $r_{\mathcal{U},x}$ is continuous and since the mapping

$$\mathbb{T}' \ni t' \mapsto X_{t'} \in \Gamma^{\nu}(\mathsf{T}\mathcal{U})$$

is Bochner integrable by hypothesis, $t \mapsto r_{\mathcal{U},x} \circ X_{t'}$ is also Bochner integrable, since Bochner integrability is preserved by continuous linear maps [1, Lemma 1.2]. Since $\mathscr{X}^x(t') = r_{\mathcal{U},x} \circ X_{t'}$ for $t' \in \mathbb{T}'$, we conclude that $\mathscr{X}$ is a locally integrally bounded local section of class C^ν.

Next suppose that $\mathscr{X}$ is a locally integrally bounded local section of class C^ν. Let $\mathbb{T}' \subseteq \mathbb{T}$ be a relatively open interval and let $\mathcal{U} \subseteq \mathsf{M}$ be open and such that $\mathbb{T}' \times \mathcal{U} \subseteq \mathcal{W}$, and let $X \in \mathrm{Map}(\mathbb{T}'; \Gamma^\nu(\mathsf{T}\mathcal{U}))$ be such that $\mathscr{X}(t, x) = [X_t]_x$. Let p be a continuous seminorm for $\Gamma^\nu(\mathsf{T}\mathcal{U})$. By Lemma 4.12 there exist $x_1, \dots, x_k \in \mathcal{U}$ and continuous seminorms q_j for $\mathscr{G}^\nu_{x_j,\mathsf{TM}}$, $j \in \{1, \dots, k\}$, such that

$$p(Y) \leq q_1 \circ r_{\mathcal{U},x_1}(Y) + \cdots + q_k \circ r_{\mathcal{U},x_k}(Y)$$

for every $Y \in \Gamma^\nu(\mathsf{T}\mathcal{U})$. By hypothesis and by definition of Bochner integrability,

$$\int_{\mathbb{T}'} q_j \circ r_{\mathcal{U},x_j}(X_t)\,\mathrm{d}x < \infty, \qquad j \in \{1, \dots, k\},$$

and so

$$\int_{\mathbb{T}'} p(X_t)\mathrm{d}t < \infty.$$

Therefore, by Lemma 3.10, $X \in \mathrm{L}^1(\mathbb{T}'; \Gamma^\nu(\mathsf{TM}))$, and so $\hat{\mathscr{X}} \in \mathrm{LI}\mathscr{G}^\nu(\mathbb{T}; \mathsf{TM})(\mathcal{W})$.

(iii) If $\hat{\mathscr{X}} \in \mathrm{LB}\mathscr{G}^\nu(\mathbb{T}; \mathsf{TM})(\mathcal{W})$ then we can deduce that $\mathscr{X}$ is a locally essentially bounded local section of class C^ν as in the corresponding part of the proof of part (ii), but using the fact that boundedness is preserved by continuous linear maps [4, §III.1.1]. Next suppose that $\mathscr{X}$ is a locally essentially bounded local section of class C^ν. Let $\mathbb{T}'$ be an open interval and let $\mathcal{U} \subseteq \mathsf{M}$ be open and such that $\mathbb{T}' \times \mathcal{U} \subseteq \mathcal{W}$, and let $X \in \mathrm{Map}(\mathbb{T}'; \Gamma^\nu(\mathsf{T}\mathcal{U}))$ be such that $\mathscr{X}(t, x) = [X_t]_x$. Let p be a continuous seminorm for $\Gamma^\nu(\mathsf{T}\mathcal{U})$. Just as in the previous part of the proof, we have $x_1, \dots, x_k \in \mathcal{U}$ and continuous seminorms q_j for $\mathscr{G}^\nu_{x_j,\mathsf{TM}}$, $j \in \{1, \dots, k\}$, such that

$$p(Y) \leq q_1 \circ r_{\mathcal{U},x_1}(Y) + \cdots + q_k \circ r_{\mathcal{U},x_k}(Y)$$

for every $Y \in \Gamma^\nu(\mathsf{T}\mathcal{U})$. By hypothesis and by Lemma 3.10, there exist $C_1, \dots, C_k \in \mathbb{R}_{>0}$ such that

$$\lambda(\{t \in \mathbb{T}' \mid q_j \circ r_{\mathcal{U},x_j}(X_t) < C_j\}) = 0,$$

and so

$$\lambda(\{t \in \mathbb{T}' \mid p(X_t) < C_1 + \cdots + C_k < \infty\}) = 0,$$

implying that $X \in \mathrm{L}^\infty(\mathbb{T}'; \Gamma^\nu(\mathsf{T}\mathcal{U}))$, and so $\hat{\mathscr{X}} \in \mathrm{LB}\mathscr{G}^\nu(\mathbb{T}; \mathsf{TM})(\mathcal{W})$. □

6.3 Étalé Open-Loop Systems and Open-Loop Subfamilies

The use of sheaves of time-varying vector fields allows us to broaden our notion of trajectory. First we need to broaden our notion of an open-loop system.

Definition 6.6 (Étalé open-loop system). Let $m \in \mathbb{Z}_{\geq 0}$ and $m' \in \{0, \text{lip}\}$, let $\nu \in \{m + m', \infty, \omega\}$, and let $r \in \{\infty, \omega\}$, as required. Let $\mathbb{T} \subseteq \mathbb{R}$ be an interval and let $\mathfrak{G} = (\mathsf{M}, \mathscr{F})$ be a C^ν-tautological control system. By $\mathrm{LI}\mathscr{G}^\nu(\mathbb{T}; \mathscr{F})$ we denote the sheaf defined by requiring that

$$\mathrm{LI}\mathscr{G}^\nu(\mathbb{T}; \mathscr{F})(\mathbb{T}' \times \mathcal{U}) = \mathrm{L}^1(\mathbb{T}'; \mathscr{F}(\mathcal{U}))$$

for every relatively open interval $\mathbb{T}' \subseteq \mathbb{T}$ and every open set $\mathcal{U} \subseteq \mathsf{M}$. An ***étalé open-loop system*** is a local section of $\mathrm{LI}\mathscr{G}^\nu(\mathbb{T}; \mathscr{F})$. ○

One can also adapt the notion of an open-loop subfamily to the étalé setting as follows.

Definition 6.7 (Étalé open-loop subfamily). Let $m \in \mathbb{Z}_{\geq 0}$ and $m' \in \{0, \text{lip}\}$, let $\nu \in \{m+m', \infty, \omega\}$, and let $r \in \{\infty, \omega\}$, as required. Let $\mathbb{T} \subseteq \mathbb{R}$ be an interval and let $\mathfrak{G} = (\mathsf{M}, \mathscr{F})$ be a C^ν-tautological control system. An ***étalé open-loop subfamily*** for $\mathfrak{G}$ is an assignment, to each open set $\mathcal{W} \subseteq \mathbb{T} \times \mathsf{M}$, a subset $\mathscr{O}_{\mathfrak{G}}(\mathcal{W}) \subseteq \mathrm{LI}\mathscr{G}^\nu(\mathbb{T}; \mathscr{F})(\mathcal{W})$ with the property that, if $\mathcal{W}_1 \subseteq \mathcal{W}_2$, then

$$\{r_{\mathcal{W}_2, \mathcal{W}_1}(X) \mid X \in \mathscr{O}_{\mathfrak{G}}(\mathcal{W}_2)\} \subseteq \mathscr{O}_{\mathfrak{G}}(\mathcal{W}_1). \quad ○$$

We can define the host of natural étalé open-loop subfamilies corresponding to those of Example 5.10.

Examples 6.8 (Étalé open-loop subfamilies). Let $m \in \mathbb{Z}_{\geq 0}$ and $m' \in \{0, \text{lip}\}$, let $\nu \in \{m + m', \infty, \omega\}$, and let $r \in \{\infty, \omega\}$, as required. Let $\mathbb{T} \subseteq \mathbb{R}$ be an interval and let $\mathfrak{G} = (\mathsf{M}, \mathscr{F})$ be a C^ν-tautological control system.

1. The ***full étalé subfamily*** for $\mathfrak{G}$ is the open-loop subfamily $\mathscr{O}_{\mathfrak{G},\text{full}}$ defined by

$$\mathscr{O}_{\mathfrak{G},\text{full}}(\mathcal{W}) = \mathrm{LI}\mathscr{G}^\nu(\mathbb{T}; \mathscr{F})(\mathcal{W}).$$

2. The ***locally essentially bounded étalé subfamily*** for $\mathfrak{G}$ is the open-loop subfamily $\mathscr{O}_{\mathfrak{G},\infty}$ defined by requiring that

$$\mathscr{O}_{\mathfrak{G},\infty}(\mathbb{T}' \times \mathcal{U}) = \{X \in \mathscr{O}_{\mathfrak{G},\text{full}}(\mathbb{T}' \times \mathcal{U}) \mid X \in \mathrm{L}^\infty(\mathbb{T}'; \Gamma^\nu(\mathsf{T}\mathcal{U}))\}$$

for every relatively open interval $\mathbb{T}' \subseteq \mathbb{T}$ and every open set $\mathcal{U} \subseteq \mathsf{M}$.
3. The ***locally essentially compact étalé subfamily*** for $\mathfrak{G}$ is the open-loop subfamily $\mathscr{O}_{\mathfrak{G},\text{cpt}}$ defined by requiring that

$$\mathscr{O}_{\mathfrak{G},\mathrm{cpt}}(\mathbb{T}' \times \mathcal{U}) = \{X \in \mathscr{O}_{\mathfrak{G},\mathrm{full}}(\mathbb{T}' \times \mathcal{U}) \mid X \in \mathrm{L}^{\mathrm{cpt}}(\mathbb{T}'; \Gamma^{\nu}(\mathsf{T}\mathcal{U}))\}$$

for every relatively open interval $\mathbb{T}' \subseteq \mathbb{T}$ and every open set $\mathcal{U} \subseteq \mathsf{M}$.

4. Next we define the ***piecewise constant étalé subfamily*** for $\mathfrak{G}$. This takes a little more work. If $\mathbb{T}' \subseteq \mathbb{T}$ is a bounded relatively open interval and $\mathcal{U} \subseteq \mathsf{M}$ is open, a map $\mathscr{X} : \mathbb{T}' \times \mathcal{U} \to \mathrm{Et}(\mathscr{F})$ is ***piecewise constant*** if there exist a finite partition $(\mathbb{T}_1, \dots, \mathbb{T}_k)$ of $\mathbb{T}'$ into pairwise disjoint intervals and vector fields $X_1, \dots, X_k \in \Gamma^{\nu}(\mathsf{T}\mathcal{U})$ such that $\mathscr{X}(t, x) = [X_j]_x$ for $t \in \mathbb{T}_j$, $j \in \{1, \dots, k\}$. Now suppose that $\mathcal{W} \subseteq \mathbb{T} \times \mathsf{M}$ is open. We say that $\mathscr{X} : \mathcal{W} \to \mathrm{Et}(\mathscr{F})$ is ***piecewise constant*** if, for every $(t, x) \in \mathcal{W}$, there exist a relatively open interval $\mathbb{T}_t \subseteq \mathbb{T}$ and an open set $\mathcal{U}_x \subseteq \mathsf{M}$ such that $(t, x) \in \mathbb{T}_t \times \mathcal{U}_x \subseteq \mathcal{W}$ and $\mathscr{X} | \mathbb{T}_t \times \mathcal{U}_x$ is piecewise constant.
5. We can associate an étalé open-loop subfamily to an étalé open-loop system as follows. Let $\mathscr{O}_{\mathfrak{G}}$ be an étalé open-loop subfamily for $\mathfrak{G}$, let $\mathcal{W} \subseteq \mathbb{T} \times \mathsf{M}$ be open, and let $X \in \mathscr{O}_{\mathfrak{G}}(\mathcal{W})$. We denote by $\mathscr{O}_{\mathfrak{G},X}$ the open-loop subfamily defined as follows. If $\mathcal{W}' \subseteq \mathcal{W}$, then we let

$$\mathscr{O}_{\mathfrak{G},X}(\mathcal{W}') = \{X' \in \mathscr{O}_{\mathfrak{G}}(\mathcal{W}') \mid X' = r_{\mathcal{W},\mathcal{W}'}(X)\}.$$

If $\mathcal{W}' \not\subseteq \mathcal{W}$, then we take $\mathscr{O}_{\mathfrak{G},X}(\mathcal{W}') = \emptyset$. ○

6.4 Étalé Trajectories

We can now define an appropriately extended notion of trajectory.

Definition 6.9 (Étalé trajectory). Let $m \in \mathbb{Z}_{\geq 0}$ and $m' \in \{0, \mathrm{lip}\}$, let $\nu \in \{m + m', \infty, \omega\}$, and let $r \in \{\infty, \omega\}$, as required. Let $\mathbb{T} \subseteq \mathbb{R}$ be an interval, let $\mathfrak{G} = (\mathsf{M}, \mathscr{F})$ be a C^{ν}-tautological control system, and let $\mathscr{O}_{\mathfrak{G}}$ be an étalé open-loop subfamily for $\mathfrak{G}$.

(i) For an open set $\mathcal{W} \subseteq \mathbb{T} \times \mathsf{M}$ and for $X \in \mathscr{O}_{\mathfrak{G}}(\mathcal{W})$, an ***étalé*** $(X, \mathcal{W})$***-trajectory*** for $\mathscr{O}_{\mathfrak{G}}$ is a curve $\xi : \mathbb{T}' \to \mathsf{M}$ such that

 (a) $\mathbb{T}' \subseteq \mathbb{T}$ is an interval,
 (b) $\mathrm{graph}(\xi) \subseteq \mathcal{W}$, and
 (c) $\xi'(t) = \mathrm{ev}_{(t,\xi(t))}(X(t, \xi(t)))$ for almost every $t \in \mathbb{T}'$.

(ii) For an open set $\mathcal{W} \subseteq \mathbb{T} \times \mathsf{M}$, an ***étalé*** $\mathcal{W}$***-trajectory*** for $\mathscr{O}_{\mathfrak{G}}$ is a curve $\xi : \mathbb{T}' \to \mathsf{M}$ such that, for some $X \in \mathscr{O}_{\mathfrak{G}}(\mathcal{W})$,

 (a) $\mathbb{T}' \subseteq \mathbb{T}$ is an interval,
 (b) $\mathrm{graph}(\xi) \subseteq \mathcal{W}$, and
 (c) $\xi'(t) = \mathrm{ev}_{(t,\xi(t))}(X(t, \xi(t)))$ for almost every $t \in \mathbb{T}'$.

(iii) An ***étalé trajectory*** for $\mathscr{O}_{\mathfrak{G}}$ is a curve that is an étalé $\mathcal{W}$-trajectory for $\mathscr{O}_{\mathfrak{G}}$ for some open set $\mathcal{W} \subseteq \mathbb{T} \times \mathsf{M}$.

We denote by:

(iv) $\text{Étraj}(X, \mathcal{W}, \mathscr{O}_{\mathfrak{G}})$ the set of étalé $(X, \mathcal{W})$-trajectories for $\mathscr{O}_{\mathfrak{G}}$;
(v) $\text{Étraj}(\mathcal{W}, \mathscr{O}_{\mathfrak{G}})$ the set of étalé $\mathcal{W}$-trajectories for $\mathscr{O}_{\mathfrak{G}}$;
(vi) $\text{Étraj}(\mathscr{O}_{\mathfrak{G}})$ the set of étalé trajectories for $\mathscr{O}_{\mathfrak{G}}$.

We shall abbreviate $\text{Étraj}(\mathcal{W}, \mathfrak{G}) = \text{Étraj}(\mathcal{W}, \mathscr{O}_{\mathfrak{G},\text{full}})$ and $\text{Étraj}(\mathfrak{G}) = \text{Étraj}(\mathscr{O}_{\mathfrak{G},\text{full}})$. ○

Let us verify that this extended notion of trajectory captures the desired behaviour from our introductory example above.

Example 6.10 (*Example 6.1 cont'd*). We claim that the curve ξ considered in Example 6.1 is an étalé trajectory. To see this, we take

$$\mathcal{W} = \underbrace{(\overbrace{(-1,3)}^{\mathbb{T}_1} \times \overbrace{((-1,3)\times(-2,0))}^{\mathcal{U}_1})}_{\mathcal{W}_1} \cup \underbrace{(\overbrace{(1,5)}^{\mathbb{T}_2} \times \overbrace{((-1,1)\times(-2,2))}^{\mathcal{U}_2})}_{\mathcal{W}_2} \cup \underbrace{(\overbrace{(3,7)}^{\mathbb{T}_3} \times \overbrace{((-1,3)\times(0,2))}^{\mathcal{U}_3})}_{\mathcal{W}_3}.$$

To define a local section of $\mathrm{LI}\mathscr{G}^{\omega}([0,6]; \mathsf{TM})$ over $\mathcal{W}$, we think of a local section as being a map from $\mathcal{W}$ into $\mathrm{Et}(\mathrm{LI}\mathscr{G}^{\omega}([0,6]; \mathsf{TM}))$, cf. Theorem 6.5. In doing so, we shall think of $\pm X$ and $\pm Y$ as being time-varying vector fields (that are actually independent of time). With this as preamble, we define $V \in \mathrm{LI}\mathscr{G}^{\omega}([0,6]; \mathscr{F}(\mathcal{W}))$ by asking that

$$V(t,x) = \begin{cases} [-X]_{(t,x)}, & (t,x) \in \mathcal{W}_1, \\ [Y]_{(t,x)}, & (t,x) \in \mathcal{W}_2, \\ [X]_{(t,x)}, & (t,x) \in \mathcal{W}_3. \end{cases}$$

One can then directly see that ξ is a $(V, \mathcal{W})$-trajectory, as desired. ○

References

1. Beckmann R, Deitmar A (2011) Strong vector valued integrals. ArXiv:1102.1246v1 [math.FA]. http://arxiv.org/abs/1102.1246v1
2. Jafarpour S, Lewis AD (2014) Time-varying vector fields and their flows. To appear in Springer Briefs in Mathematics
3. MacLane S, Moerdijk I (1992) Sheaves in geometry and logic: a first introduction to topos theory. Universitext. Springer, New York/Heidelberg/Berlin
4. Schaefer HH, Wolff MP (1999) Topological Vector Spaces, 2 edn. No. 3 in Graduate Texts in Mathematics. Springer, Berlin
5. Stacks Project Authors (2014) Stacks Project. http://stacks.math.columbia.edu

Chapter 7
Ongoing and Future Work

In our discussion of tautological control systems in Chap. 5, we strove to make connections between tautological control systems and standard control models. We do not wish to give the impression, however, that tautological control systems are mere fancy reformulations of standard control systems. In this chapter we will give a sketchy, but we hope compelling, idea of how the tautological control system framework can be used to say new things about control systems. This will also provide an illustration of how, in practice, one can do control theory within the confines of the tautological control system framework, without reverting to the comforting control parameterisations with which one is familiar. We will emphasise that some of these ideas are in the preliminary stages of investigation, so the final word on what results will look like has yet to be uttered. Nonetheless, we believe that even the clear problem formulations we give make it apparent that there is something "going on" here.

7.1 Linearisation

Work has already been completed on a theory for linearisation of tautological control systems [36]. The development of even this elementary control theoretic construction takes some nontrivial effort. However, it does provide one with some insights into linearisation, and here we will give a brief outline of the main ideas.

We let $m \in \mathbb{Z}_{>0}$, $m' \in \{0, \mathrm{lip}\}$, $\nu \in \{m + m', \infty, \omega\}$, and $r \in \{\infty, \omega\}$, as appropriate. We let $\mathfrak{G} = (\mathsf{M}, \mathscr{F})$ be a C^ν-tautological control system. The first step in the process is to define the linearisation of the *system*. This is itself a tautological control system $T\mathfrak{G}$ with the tangent bundle TM as state space. The presheaf $T\mathscr{F}$ of sets of vector fields is given by requiring that

$$T\mathscr{F}(\pi_{\mathsf{TM}}^{-1}(\mathcal{U})) = \{X^T + Y^V \mid X, Y \in \mathscr{F}(\mathcal{U})\}$$

A.D. Lewis, *Tautological Control Systems*, SpringerBriefs in Control, Automation and Robotics, DOI: 10.1007/978-3-319-08638-5_7

for every open $\mathcal{U} \subseteq \mathsf{M}$. Here X^T is the ***tangent lift*** of X, which is defined by

$$X^T(v_x) = \left.\frac{\mathrm{d}}{\mathrm{d}t}\right|_{t=0} T_x\Phi_t^X(v_x),$$

and Y^V is the ***vertical lift*** of Y, which is defined by

$$Y^V(v_x) = \left.\frac{\mathrm{d}}{\mathrm{d}t}\right|_{t=0} (v_x + tY(x)).$$

This definition of linearisation can be given a quite precise motivation, and this is explained in [36, Eq. (14)]. For our purposes here, we shall simply note that X^T can be thought of as the linearisation with respect to state and Y^V can be thought of as linearisation with respect to control. One can show that the linearisation of a C^ν-tautological control system is a $\mathrm{C}^{\nu-1}$-tautological control system, with the obvious conventions $\infty - 1 = \infty$ and $\omega - 1 = \omega$.

One then needs to define trajectories for the linearisation $T\mathfrak{G}$. This becomes a nontrivial task. The first step is that one needs to properly extend tangent and vertical lifts to time-varying vector fields from the class $\mathrm{LI}\Gamma^\nu(\mathbb{T}; \mathsf{TTM})$. For the vertical lift, this is done easily, but the extension of the tangent lift construction is more complicated. One anticipates, and it is true, that, if $X \in \mathrm{LI}\Gamma^\nu(\mathbb{T}; \mathsf{TM})$, then the proper definition should give $X^T \in \mathrm{LI}\Gamma^{\nu-1}(\mathbb{T}; \mathsf{TTM})$. The work required to show this is done in [36]. This then allows one to define appropriate open-loop systems, open-loop families, and then trajectories for the linearisation. As one expects, a trajectory for the linearisation is a vector field along a trajectory for the system itself.

One of the issues that the process of linearisation for tautological control systems brings into focus is the distinction between linearisation along a trajectory and linearisation along a reference flow. Associated to a reference flow, i.e., a prescribed open-loop system, and an initial condition will be a unique trajectory. However, a trajectory can correspond to multiple reference flows. This allowing for multiple reference flows when linearising about a trajectory shows up in a pointed way when considering equilibrium trajectories, i.e., stationary trajectories. In this case, it is natural when linearising in the tautological framework to have a *time-varying* linearisation of a system about an equilibrium point. This is something that is simply missing from classical Jacobian linearisation.

Having described in broad terms the process by which one builds the linearisation of a tautological control system, let us indicate, without the details which can be found in [36], how the constructions resolve the potential confusion of Example 1.1. If one regards this system as a tautological control system with equilibrium point at $(0, 0, 0)$, then the linearisation about the equilibrium trajectory $t \mapsto (0, 0, 0)$ is time-varying in this case. However, one notices that the equilibrium trajectory can be a trajectory for an infinite number of open-loop systems. By linearising about one or the other of these, one will get linearisations with different controllability properties. This makes one think about what should be the proper definition of linear controllability in this case, and this is done in [36] with an invariant subspace characterisation, and the tautological control system of Example 1.1 is shown to be linearly controllable according to this correct feedback-invariant definition.

For our purposes the central point to take from this discussion is that, if one strictly adheres to the tautological control system framework, extra structure—in this case time-varying linearisations of time-independent systems about equilibria—reveals itself.

7.2 Optimal Control Theory

Ongoing work, presently not complete, involves formulating problems in optimal control theory in the tautological control system framework. It is possible, at this point, to state the basic ingredients of the formulation, but the exact statement of the results and their proofs will require additional work.

We let $m \in \mathbb{Z}_{>0}$, $m' \in \{0, \text{lip}\}$, $\nu \in \{m + m', \infty, \omega\}$, and $r \in \{\infty, \omega\}$, as appropriate. By $\mathscr{C}^\nu_{\mathsf{M}}$ we denote the sheaf of C^ν-functions, this being defined in exactly the same manner as the sheaf $\mathscr{G}^\nu_{\mathsf{TM}}$, with "vector field" replaced by "function". We let $\mathfrak{G} = (\mathsf{M}, \mathscr{F})$ be a C^ν-tautological control system. A C^ν***-Lagrangian*** for $\mathfrak{G}$ is a family of mappings $\Lambda_{\mathcal{U}}\colon \mathscr{F}(\mathcal{U}) \to \mathscr{C}^\nu_{\mathsf{M}}(\mathcal{U})$, $\mathcal{U} \subseteq \mathsf{M}$ open, with the following properties:

1. if $\mathcal{U}, \mathcal{V} \subseteq \mathsf{M}$ are open with $\mathcal{V} \subseteq \mathcal{U}$, then $\Lambda_{\mathcal{V}}(X|\mathcal{V}) = \Lambda_{\mathcal{U}}(X)|\mathcal{V}$, i.e., "$\Lambda$ commutes with restriction";
2. $\Lambda_{\mathcal{U}}$ is continuous with respect to the C^ν-topologies for $\mathscr{F}(\mathcal{U})$ and $\mathscr{C}^\nu_{\mathsf{M}}(\mathcal{U})$.

Note that a Lagrangian is no longer a function, as in "ordinary" optimal control theory, but is actually a morphism of presheaves of locally convex topological vector spaces [41, Definition 1.1.9]. Let us see how the usual notion of a Lagrangian in optimal control theory is related to the preceding definition, as this is not immediately obvious. Thus we suppose that we have a C^ν-control system $\Sigma = (\mathsf{M}, F, \mathcal{C})$ with the property that $\hat{F}\colon \mathcal{C} \to \Gamma^\nu(\mathsf{TM})$ is continuous, open, and injective, where $\hat{F}(u) = F(x, u)$. It is reasonable to suppose a Lagrangian that shares the regularity of Σ. Thus we suppose that $L\colon \mathsf{M} \times \mathcal{C} \to \mathbb{R}$ is such that the map $\hat{L}\colon \mathcal{C} \to \mathrm{C}^\nu(\mathsf{M})$ is continuous, if $\mathrm{C}^\nu(\mathsf{M})$ has the C^ν-topology and $\hat{L}(u) = L(x, u)$. This, obviously, echoes the way that we prescribe regularity for control systems, and we refer to [26, §4] for a more concrete description. A Lagrangian L for a control system Σ defines a Lagrangian Λ_L for the corresponding tautological control system $\mathfrak{G}_\Sigma$ by

$$\Lambda_L(F^u)(x) = L(x, u).$$

Since $\Lambda_L \circ \hat{F} = \hat{L}$ and since $\hat{F}$ is continuous, open, and injective, it follows that Λ_L is indeed continuous.

Now, given a Lagrangian for a tautological control system, one can define an associated ***Hamiltonian*** as the family of mappings $H_{\mathfrak{G},\Lambda,\mathcal{U}}\colon \mathscr{F}(\mathcal{U}) \to \mathscr{C}^\nu_{\mathsf{T}^*\mathsf{M}}(\mathsf{T}^*\mathcal{U})$, $\mathcal{U} \subseteq \mathsf{M}$ open, defined by

$$H_{\mathfrak{G},\Lambda,\mathcal{U}}(X)(\alpha_x) = \langle \alpha_x; X(x) \rangle + \Lambda_{\mathcal{U}}(X)(x).$$

Thus an Hamiltonian is not a function on the cotangent bundle, but rather again a sort of presheaf morphism (in a way we will not make precise here). Associated with this Hamiltonian is an ***Hamiltonian vector field*** which is the family of mappings $\mathscr{H}_{\mathfrak{G},\Lambda,\mathcal{U}}\colon \mathscr{F}(\mathcal{U}) \rightarrow \mathscr{G}^{\nu-1}_{\mathsf{T}^*\mathsf{M}}(\mathsf{T}^*\mathcal{U})$, $\mathcal{U} \subseteq \mathsf{M}$ open, defined by

$$\mathscr{H}_{\mathfrak{G},\Lambda,\mathcal{U}}(X) = \omega^{\sharp} \circ \mathrm{d}(H_{\mathfrak{G},\Lambda,\mathcal{U}}(X)),$$

where ω is the canonical symplectic form on $\mathsf{T}^*\mathsf{M}$. As with our other constructions, we see that, in the tautological control system framework, this is not a vector field, but some kind of presheaf morphism.

We can see, then, that optimal control theory in the tautological control system setting is not "just the same" as in the ordinary setting, and one expects the presheaf structure that runs through the above constructions to be important in the development of the theory, although there are still many details left to work out.

We can make the preceding constructions concrete by considering a specific application in optimal control theory, that of sub-Riemannian geometry. The presentation we give, in the smooth case, is shared with the presentation of sub-Riemannian geometry by Sussmann in [51]. It is worth noting, however, that Sussmann's approach does not immediately extend to the real analytic case, as the topology on the space of real analytic vector fields is not Fréchet.

Let us define the basic structure of sub-Riemannian geometry.

Definition 7.1 (Sub-Riemannian manifold). Let $m \in \mathbb{Z}_{\geq 0}$ and $m' \in \{0, \mathrm{lip}\}$, let $\nu \in \{m + m', \infty, \omega\}$, and let $r \in \{\infty, \omega\}$, as required. A ***C$^{\nu}$-sub-Riemannian manifold*** is a pair $(\mathsf{M}, \mathbb{G})$ where M is a C^r-manifold and $\mathbb{G}$ is a C^{ν}-tensor field of type $(2, 0)$ such that, for each $x \in \mathsf{M}$, $\mathbb{G}(x)$ is positive-semidefinite as a quadratic function on $\mathsf{T}^*_x\mathsf{M}$. ∘

Associated with a sub-Riemannian structure $\mathbb{G}$ on M is a distribution that we now describe. First of all, we have a map $\mathbb{G}^{\sharp}\colon \mathsf{T}^*\mathsf{M} \rightarrow \mathsf{TM}$ defined by

$$\langle \beta_x; \mathbb{G}^{\sharp}(\alpha_x)\rangle = \mathbb{G}(\beta_x, \alpha_x).$$

We then denote by $\mathsf{D}_{\mathbb{G}} = \mathrm{image}(\mathbb{G}^{\sharp})$ the associated distribution. Note that $\mathsf{D}_{\mathbb{G}}$ is a distribution of class C^{ν} since, for each $x \in \mathsf{M}$, there exist a neighbourhood $\mathcal{U}$ of x and a family of C^{ν}-vector fields $(X_a)_{a \in A}$ on $\mathcal{U}$ (namely the images under $\mathbb{G}^{\sharp}$ of the coordinate basis vector fields, if we choose $\mathcal{U}$ to be a coordinate chart domain) such that

$$\mathsf{D}_{\mathbb{G},y} = \mathsf{D}_{\mathbb{G}} \cap \mathsf{T}_y\mathsf{M} = \mathrm{span}_{\mathbb{R}}(X_a(y)|\ a \in A)$$

for every $y \in \mathcal{U}$. There is also an associated ***sub-Riemannian metric*** for $\mathsf{D}_{\mathbb{G}}$, i.e., an assignment to each $x \in \mathsf{M}$ an inner product $\mathbb{G}(x)$ on $\mathsf{D}_{\mathbb{G},x}$. This is denoted also by $\mathbb{G}$ and defined by

$$\mathbb{G}(u_x, v_x) = \mathbb{G}(\alpha_x, \beta_x),$$

where $u_x = \mathbb{G}^\sharp(\alpha_x)$ and $v_x = \mathbb{G}^\sharp(\beta_x)$, and where we joyously abuse notation.

An absolutely continuous curve $\gamma\colon [a,b] \to \mathsf{M}$ is $\mathsf{D}_{\mathbb{G}}$***-admissible*** if $\gamma'(t) \in \mathsf{D}_{\mathbb{G},\gamma(t)}$ for almost every $t \in [a,b]$. The ***length*** of a $\mathsf{D}_{\mathbb{G}}$-admissible curve $\gamma\colon [a,b] \to \mathsf{M}$ is

$$\ell_{\mathbb{G}}(\gamma) = \int_a^b \sqrt{\mathbb{G}(\gamma'(t), \gamma'(t))}\, \mathrm{d}t.$$

As in Riemannian geometry, the length of a $\mathsf{D}_{\mathbb{G}}$-admissible curve is independent of parameterisation, and so curves can be considered to be defined on $[0,1]$. We can then define the ***sub-Riemannian distance*** between $x_1, x_2 \in \mathsf{M}$ by

$$\mathrm{d}_{\mathbb{G}}(x_1, x_2) = \inf\{\ell_{\mathbb{G}}(\gamma) \mid \gamma\colon [0,1] \to \mathsf{M} \text{ is an absolutely continuous curve for which } \gamma(0) = x_1 \text{ and } \gamma(1) = x_2\}.$$

One of the problems of sub-Riemannian geometry is to determine length minimising curves, i.e., sub-Riemannian geodesics.

A common means of converting sub-Riemannian geometry into a standard control problem is to choose a $\mathbb{G}$-orthonormal basis $(X_1, \dots, X_k)$ for $\mathsf{D}_{\mathbb{G}}$ and so consider the control-affine system with dynamics prescribed by

$$F(x, \mathbf{u}) = \sum_{a=1}^{k} u^a X_a(x), \qquad x \in \mathsf{M},\ \mathbf{u} \in \mathbb{R}^k.$$

Upon doing this, $\mathsf{D}_{\mathbb{G}}$-admissible curves are evidently trajectories for this control-affine system. Moreover, for a trajectory $\xi\colon [0,1] \to \mathsf{M}$ satisfying

$$\xi'(t) = \sum_{a=1}^{k} u^a(t) X_a(\xi(t)),$$

we have

$$\ell_{\mathbb{G}}(\xi) = \int_0^1 \|\mathbf{u}(t)\|\, \mathrm{d}t.$$

The difficulty, of course, with the preceding approach to sub-Riemannian geometry is that there may be no $\mathbb{G}$-orthonormal basis for $\mathsf{D}_{\mathbb{G}}$. This can be the case for at least two reasons: (1) the distribution $\mathsf{D}_{\mathbb{G}}$ may not have locally constant rank; (2) when the distribution $\mathsf{D}_{\mathbb{G}}$ has locally constant rank, the global topology of M may prohibit the existence of a global basis, e.g., on even-dimensional spheres there is no global basis for vector fields, orthonormal or otherwise. However, one can formulate sub-Riemannian geometry in terms of a tautological control system in a natural way. Indeed, associated to $\mathsf{D}_{\mathbb{G}}$ is the tautological control system $\mathfrak{S}_{\mathbb{G}} = (\mathsf{M}, \mathscr{F}_{\mathbb{G}})$, where, for an open subset $\mathcal{U} \subset \mathsf{M}$,

$$\mathscr{F}_{\mathbb{G}}(\mathcal{U}) = \{X \in \Gamma^{\nu}(\mathsf{T}\mathcal{U}) \mid X(x) \in \mathsf{D}_{\mathbb{G},x},\ x \in \mathcal{U}\}.$$

One readily verifies that $\mathscr{F}_{\mathbb{G}}$ is a sheaf. Note that the sheaf $\mathscr{F}_{\mathbb{G}}$ is not globally generated; this is because it is a sheaf, cf. Example 4.3–3. However, it can be regarded as the sheafification of the globally generated sheaf with global generators $\mathscr{F}_{\mathbb{G}}(\mathsf{M})$.

Lemma 7.2 (The sheaf of vector fields for the sub-Riemannian tautological control problem). *The sheaf $\mathscr{F}_{\mathbb{G}}$ is the sheafification of the globally generated presheaf with generators $\mathscr{F}_{\mathbb{G}}(\mathsf{M})$.*

Proof This is a result about sheaf cohomology, and we will not give all details here. Instead we will simply point to the main facts from which the conclusion follows. First of all, to prove the assertion, it suffices by Lemma 4.6 to show that $\mathscr{F}_{\mathbb{G},x}$ is generated, as a module over the ring $\mathscr{C}^{\nu}_{x,\mathsf{M}}$, by germs of global sections. In the cases $\nu \in \{m, m + \mathrm{lip}, \infty\}$, the fact that the sheaf of rings of smooth functions admits partitions of unity implies that the sheaf $\mathscr{C}^{\nu}_{\mathsf{M}}$ is a fine sheaf of rings [54, Example 3.4(d)]. It then follows from [54, Example 3.4(e)] that the sheaf $\mathscr{F}_{\mathbb{G}}$ is also fine and so soft [54, Proposition 3.5]. Because of this, the cohomology groups of positive degree for this sheaf vanish [54, Proposition 3.11], and this ensures that germs of global sections generate all stalks (more or less by definition of cohomology in degree 1). In the case $\nu = \omega$, the result is quite nontrivial. First of all, by a real analytic adaptation of [16, Corollary H9], one can show that $\mathscr{F}_{\mathbb{G}}$ is locally finitely generated. Then, $\mathscr{F}_{\mathbb{G}}$ being a finitely generated subsheaf of the coherent sheaf $\mathscr{G}^{\omega}_{\mathsf{TM}}$, it is itself coherent [14, Theorem 3.16]. Then, by Cartan's Theorem A [11], we conclude that $\mathscr{F}_{\mathbb{G},x}$ is generated by germs of global sections. □

Let us next formulate the sub-Riemannian geodesic problem in the framework of tautological control systems. First of all, it is convenient when performing computations to work with energy rather than length as the quantity we are minimising. To this end, for an absolutely continuous $\mathsf{D}_{\mathbb{G}}$-admissible curve $\gamma\colon [a, b] \to \mathsf{M}$, we define the ***energy*** of this curve to be

$$E_{\mathbb{G}}(\gamma) = \frac{1}{2}\int_a^b \mathbb{G}(\gamma'(t), \gamma'(t))\,\mathrm{d}t.$$

A standard argument shows that curves that minimise energy are in 1–1 correspondence with curves that minimise length and are parameterised to have an appropriate constant speed [37, Proposition 1.4.3]. We can and do, therefore, consider the energy minimisation problem. In terms of our general constructions above for optimal control in the tautological control system setting, the Lagrangian for sub-Riemannian geometry will be the family $(\Lambda_{\mathbb{G},\mathcal{U}})_{\mathcal{U}\ \mathrm{open}}$ of mappings defined by

$$\Lambda_{\mathbb{G},\mathcal{U}}(X)(x) = \tfrac{1}{2}\mathbb{G}(X(x), X(x)).$$

The corresponding Hamiltonian will be the family of mappings $(H_{\mathbb{G},\mathcal{U}})_{\mathcal{U}\text{ open}}$ defined by

$$H_{\mathbb{G},\mathcal{U}}(X)(\alpha_x) = \langle \alpha_x; X(x)\rangle + \lambda_0 \tfrac{1}{2}\mathbb{G}(X(x), X(x)),$$

where $\lambda_0 \in \{0, -1\}$, with $\lambda_0 = 0$ corresponding to so-called abnormal extremals. Let us apply the classical Maximum Principle [39], leaving aside the technicalities caused by the complicated topology of the control set. If we consider only normal extremals, i.e., supposing that $\lambda_0 = -1$, then the Maximum Principle prescribes that the reference flow giving an optimal trajectory will necessarily be a bundle map $X_*\colon \mathsf{T}^*\mathsf{M} \to \mathsf{TM}$ over id_M chosen so that $X_*(\alpha_x)$ maximises the function

$$v_x \mapsto \langle \alpha_x; v_x\rangle - \tfrac{1}{2}\mathbb{G}(v_x, v_x).$$

Standard finite-dimensional optimisation gives $X_*(x) = \mathbb{G}^\sharp(\alpha_x)$. The ***maximum Hamiltonian*** is then obtained by substituting this value of the "control" into the Hamiltonian:

$$\begin{aligned} H^{\max}_{\mathbb{G}}(X_*)\colon \mathsf{T}^*\mathsf{M} &\to \mathbb{R} \\ \alpha_x &\mapsto \tfrac{1}{2}\mathbb{G}(\alpha_x, \alpha_x). \end{aligned}$$

The normal extremals are then integral curves of the Hamiltonian vector field associated with the Hamiltonian $H^{\max}_{\mathbb{G}}(X_*)$. At a superficial level and in the normal case, at least, we thus see that the tautological control system formulation of sub-Riemannian geometry gives the familiar extremals [37, Theorem 1.5.7].

The preceding computations, having banished the usual parameterisation by control, are quite elegant when compared to manner in which one applies the Maximum Principle to the "usual" control formulation of sub-Riemannian geometry. The calculations are also more general and global. However, to make sense of them, one has to prove an appropriate version of the Maximum Principle, something which will be forthcoming. For the moment, we mention that a significant rôle in this will be played by appropriate needle variations constructed by dragging variations along a trajectory to the final endpoint. The manner in which one drags these variations has to do with linearisation, as described in Sect. 7.1.

7.3 Controllability

In this section we will vaguely sketch some ideas that are beginning to emerge regarding the study of controllability in the tautological control system framework. Before we do this, let us give a brief critical overview of the state of the literature on the controllability problem, referring to [35] for a more organised and extensive version of this.

The controllability of nonlinear systems comprises a vast and difficult component of the geometric control theory literature. A number of papers have been published addressing the seemingly impenetrable nature of the problems of controllability [1, 6, 29, 30, 34, 42]. Despite this, there has been substantial effort dedicated to determining sufficient or necessary conditions for controllability [2, 4, 7–9, 17–23, 27, 28, 31–33, 46, 48–50, 52]. The problem of controllability has a certain lure that attracts researchers in geometric control theory. The problem is such a natural one that it feels as if it should be possible to obtain complete results, at least in some quite general situations. However, this objective remains to be fulfilled.

Our view is that one of the reasons for this is that many of the approaches to controllability are not feedback-invariant. An extreme example of this are methods for studying controllability of control-affine systems, *fixing* a drift vector field f_0 and control vector fields $f_1, \dots, f_m$, and using these as generators of a free Lie algebra. In this sort of analysis, Lie series are truncated, leading to the notion of "nilpotent approximation" of control systems. These ideas are reflected in a great many of the papers cited above. The difficulty with this approach is that it will behave very badly under feedback transformations, cf. Example 1.2. This is discussed in [35].

One approach is then to attempt to find feedback-invariant conditions for local controllability. The first-order case is considered in [3, Theorem 5.16]; see also [7]. Second-order feedback-invariant conditions are considered in [5, 24]. Any attempts to determine higher-order feedback-invariant controllability conditions have, as far as we know, met with no success. Indeed, the likelihood of this approach leading anywhere seems very small, given the extremely complicated manner in which feedback transformations interact with controllability conditions.

Thus controllability theory would appear to be an area where the tautological control system framework, with its "built-in" feedback-invariance, might be useful. In work with the author's Doctoral student, Saber Jafarpour, the Orbit Theorem of Stefan [45] and Sussmann [47] is being adapted to tautological control systems. The results themselves are not surprising. However, in carrying out this adaptation, two related observations have surfaced.

1. For the purposes of defining reachable sets and orbits, and, therefore, for studying controllability, the notion of an étalé trajectory as in Definition 6.9 is the correct notion to use, rather than the more straightforward construction of Definition 5.14. This should not be surprising, given Example 6.1.
2. That tautological control systems lead us naturally to étalé trajectories has a potentially profound consequence. As is clear from the constructions of Sects. 6.1 and 6.2, étalé trajectories are intimately and nontrivially connected with, not sets of vector fields, but *sheaves* of sets of vector fields, and all of the structure that this possesses and imposes, especially concerning étalé spaces and stalk topologies. When considering flows, i.e., one-parameter families of diffeomorphisms, this leads one naturally to pseudogroups and groupoids. This idea is touched upon by Stefan [45], but other than this is completely unexplored in control theory.

Thus, while this work on controllability is in its formative stages, in these formative stages are already seen new ideas for attacking important problems in control theory.

7.4 Feedback and Stabilisation Theory

There are, one could argue, three big fundamental problems in geometric control theory. Two, controllability and optimal control, are discussed above, and moreover in the context of tautological control systems. The third is stabilisation with which we have as yet done no work in the tautological control system framework. The stabilisation problem, being one of enormous practical importance, has been comprehensively studied, mainly from the point of view of Lyapunov theory, where the notion of a "control-Lyapunov function" provides a useful device for characterising when a system is stabilisable [12] and for stabilisation if one is known [43]. Our view is that Lyapunov characterisations for stabilisability are important from a practical point of view, but, from a fundamental point of view, merely replace one impenetrable notion, "stabilisability", with another, "existence of a control-Lyapunov function". This is expressed succinctly by Sontag.

> In any case, all converse Lyapunov results are purely existential, and are of no use in guiding the search for a Lyapunov function. The search for such functions is more of an art than a science, and good physical insight into a given system plus a good amount of trial and error is typically the only way to proceed.—[44, page 259]

As Sontag goes on to explain, there are many heuristics for guessing control-Lyapunov functions. However, this is unsatisfying if one is seeking a general understanding of the problem of stabilisability, and not just a means of designing stabilising controllers for individual systems or classes of systems.

It is also the case that there has been virtually no work on stabilisability from a geometric perspective. Topological characterisations of stabilisability such as those of [10] (refined in [38, 55]) and [13] are gratifying when they are applicable, but they are far too coarse to provide anything even close to a complete characterisation of the problem. Indeed, the extremely detailed and intricate analysis of controllability, as reflected by the work we cite above, is simply not present for stabilisability. It is fair to say that, outside the general control-Lyapunov function framework, very little work has been done in terms of really understanding the structural obstructions to stabilisability. Moreover, it is also fair to say that, again outside the general control-Lyapunov function framework, almost none of the published literature on stabilisation and stabilisability passes the "acid test" for feedback-invariance that we discuss in Sect. 1.1. For researchers such as ourselves interested in structure, this in an unsatisfying state of affairs.

Our framework provides a natural means of addressing problems like this, just as with controllability and optimal control, because of the feedback-invariance of the framework. Indeed, upon reflection, one sees that the problem of stabilisability should have some relationships with that of controllability, although little work has been done along these lines (but see the PhD thesis of Isaiah [25]). This area of research is wide open [35].

7.5 The Category of Tautological Control Systems

In Sect. 5.6 we introduced morphisms between tautological control systems with the objective of showing that our framework is feedback-invariant. The notion of morphism we present is one that is natural and possibly easy to work with. It would be, therefore, interesting to do all of the exercises of category theory with the category of tautological control systems. That is, one would like to study epimorphisms, monomorphisms, subobjects, quotient objects, products, coproducts, pullbacks, push-outs, and various functorial operations in this category. Many of these may not be interesting or useful, or even exist. But probably some of it would be of interest. For example, Tabuada and Pappas [53] study quotients of control systems, and Elkin [15] studies various categorical constructions for control-affine systems.

References

1. Agrachev AA (1999) Is it possible to recognize local controllability in a finite number of differentiations? Open problems in mathematical systems and control theory, Communications and Control Engineering Series. Springer, Heidelberg, pp 15–18
2. Agrachev AA, Gamkrelidze RV (1993) Local controllability and semigroups of diffeomorphisms. Acta Appl Math 32(1):1–57
3. Aguilar CO (2010) Local controllability of affine distributions. PhD thesis. Queen's University, Kingston, Kingston, ON, Canada
4. Bacciotti A, Stefani G (1983) On the relationship between global and local controllability. Math Syst Theor 16(1):79–91
5. Basto-Gonçalves J (1998) Second-order conditions for local controllability. Syst Control Lett 35(5):287–290
6. Bianchini RM, Kawski M (2003) Needle variations that cannot be summed. SIAM J Control Optim 42(1):218–238
7. Bianchini RM, Stefani G (1984) Normal local controllability of order one. Int J Control 39(4):701–714
8. Bianchini RM, Stefani G (1986) Local controllability along a reference trajectory. Analysis and optimization of systems, vol 83. Lecture Notes in Control and Information Sciences. Springer, Berlin, pp 342–353
9. Bianchini RM, Stefani G (1993) Controllability along a trajectory: a variational approach. SIAM J Control Optim 31(4):900–927
10. Brockett RW (1983) Asymptotic stability and feedback stabilization. In: Brockett RW, Millman RS, Sussmann HJ (eds) Differential geometric control theory, No. 27 in Progress in Mathematics, pp 181–191. Birkhäuser, Boston
11. Cartan H (1957) Variétés analytiques réelles et variétés analytiques complexes. Bull Soc Math France 85:77–99
12. Clarke FH, Ledyaev YS, Sontag ED, Subotin AI (1997) Asymptotic controllability implies feedback stabilization. Institute of Electrical and Electronics Engineers. Trans Automat Control 42(10):1394–1407

13. Coron JM (1990) A necessary condition for feedback stabilization. Syst Control Lett 14(3): 227–232
14. Demailly JP (2012) Complex analytic and differential geometry. Unpublished manuscript made publicly available (2012). http://www-fourier.ujf-grenoble.fr/~demailly/manuscripts/agbook.pdf
15. Elkin VI (1999) Reduction of nonlinear control systems. A differential geometric approach. No. 472 in Mathematics and its Applications. Kluwer Academic Publishers, Dordrecht (Translated from the 1997 Russian original by Naidu PSV)
16. Gunning RC (1990) Introduction to holomorphic functions of several variables, vol II: local theory. Wadsworth & Brooks/Cole Mathematics Series. Wadsworth & Brooks/Cole, Belmont
17. Haynes GW, Hermes H (1970) Nonlinear controllability via Lie theory. J Soc Indust Appl Math Series A Control 8:450–460
18. Hermes H (1974) On local and global controllability. J Soc Indust Appl Math Series A Control 12:252–261
19. Hermes H (1976) High order conditions for local controllability and controlled stability. In: Proceedings of the 1976 IEEE Conference on Decision and control. Institute of Electrical and Electronics Engineers, Clearwater, pp 836–840
20. Hermes H (1976) Local controllability and sufficient conditions in singular problems. J Differ Equ 20(1):213–232
21. Hermes H (1977) High order controlled stability and controllability. Dynamical systems. Academic Press, New York, pp 89–99 (Proceedings of International Symposium, Gainesville, FL)
22. Hermes H (1982) On local controllability. SIAM J Control Optim 20(2):211–220
23. Hermes H, Kawski M (1987) Local controllability of a single input, affine system. In: Lakshmikantham V (ed) Nonlinear analysis and applications, vol 109. Lecture Notes in Pure and Applied Mathematics. Dekker Marcel Dekker, New York, pp 235–248
24. Hirschorn RM, Lewis AD (2002) Geometric local controllability: second-order conditions. In: Proceedings of the 41st IEEE Conference on decision and control. Institute of Electrical and Electronics Engineers, Las Vegas, pp 368–369
25. Isaiah P (2012) Feedback stabilisation of locally controllable systems. PhD thesis. Queen's University, Kingston, ON, Canada
26. Jafarpour S, Lewis AD (2014) Locally convex topologies and control theory. Submitted to SIAM J Control Optim
27. Kawski M (1987) A necessary condition for local controllability. Differential geometry: the interface between pure and applied mathematics, vol 68. Contemporary Mathematics. American Mathematical Society, Providence, RI, pp 143–155
28. Kawski M (1988) Control variations with an increasing number of switchings. Bull Amer Math Soc (N.S.) 18(2):149–152
29. Kawski M (1990) The complexity of deciding controllability. Syst Control Lett 15(1):9–14
30. Kawski M (1990) High-order small-time local controllability. Nonlinear controllability and optimal control, vol 133. Monographs and Textbooks in Pure and Applied Mathematics, Dekker Marcel Dekker, New York, pp 431–467
31. Kawski M (1992) High-order conditions for local controllability in practice. Recent advances in mathematical theory of systems, control, networks and signal processing, II. Mita, Tokya, pp 271–276
32. Kawski M (1998) Nonlinear control and combinatorics of words. In: Jakubczyk B, Respondek W (eds) Geometry of feedback and optimal control. Dekker Marcel Dekker, New York, pp 305–346
33. Kawski M (1999) Controllability via chronological calculus. In: Proceedings of the 38th IEEE Conference on decision and control. Institute of Electrical and Electronics Engineers, Phoenix, AZ, pp 2920–2926
34. Kawski M (2006) On the problem whether controllability is finitely determined. In: Proceedings of MTNS '06
35. Lewis AD (2012) Fundamental problems of geometric control theory. In: Proceedings of the 51st IEEE Conference on decision and control. Institute of Electrical and Electronics Engineers, Maui, HI, pp 7511–7516

36. Lewis AD (2014) Linearisation of tautological control systems. Submitted to J Geom Mech
37. Montgomery R (20032) A tour of subriemannian geometries, their geodesics and applications, No. 91 in American Mathematical Society Mathematical Surveys and Monographs. American Mathematical Society, Providence, RI
38. Orsi R, Praly L, Mareels IMY (2003) Necessary conditions for stability and attractivity of continuous systems. Int J Control 76(11):1070–1077
39. Pontryagin LS, Boltyanskii VG, Gamkrelidze RV, Mishchenko EF (1961) Matematicheskaya teoriya optimal' nykh protsessov. Gosudarstvennoe izdatelstvo fiziko-matematicheskoi literatury, Moscow (Reprint of translation: [40])
40. Pontryagin LS, Boltyanskii VG, Gamkrelidze RV, Mishchenko EF (1986) The mathematical theory of optimal processes. Classics of Soviet Mathematics. Gordon and Breach Science Publishers, New York (Reprint of 1962 translation from the Russian by Trirogoff KN)
41. Ramanan S (2005) Global Calculus, No. 65 in Graduate Studies in Mathematics. American Mathematical Society, Providence, RI
42. Sontag ED (1988) Controllability is harder to decide than accessibility. SIAM J Control Optim 26(5):1106–1118
43. Sontag ED (1989) A "universal" construction of Artstein's theorem on nonlinear stabilization. Syst Control Lett 13(2):117–123
44. Sontag ED (1998) Mathematical control theory: deterministic finite dimensional systems, 2 edn, No. 6 in Texts in Applied Mathematics. Springer, Heidelberg
45. Stefan P (1974) Accessible sets, orbits and foliations with singularities. Proc London Math Soc Third Series 29:699–713
46. Stefani G (1986) On the local controllability of a scalar-input control system. In: Theory and applications of nonlinear control systems. North-Holland, pp 167–179
47. Sussmann HJ (1973) Orbits of families of vector fields and integrability of distributions. Trans Am Math Soc 180:171–188
48. Sussmann HJ (1978) A sufficient condition for local controllability. SIAM J Control Optim 16(5):790–802
49. Sussmann HJ (1983) Lie brackets and local controllability: a sufficient condition for scalar-input systems. SIAM J Control Optim 21(5):686–713
50. Sussmann HJ (1987) A general theorem on local controllability. SIAM J Control Optim 25(1):158–194
51. Sussmann HJ (1997) An introduction to the coordinate-free maximum principle. In: Jakubczyk B, Respondek W (eds) Geometry of feedback and optimal control. Dekker Marcel Dekker, New York, pp 463–557
52. Sussmann HJ, Jurdjevic V (1972) Controllability of nonlinear systems. J Differ Equ 12:95–116
53. Tabuada P, Pappas GJ (2005) Quotients of fully nonlinear control systems. SIAM J Control Optim 43(5):1844–1866
54. Wells RO Jr (2008) Differential analysis on complex manifolds, 3 edn, No. 65 in Graduate Texts in Mathematics. Springer, New York
55. Zabczyk J (1989) Some comments on stabilizability. Appl Math Optim 19(1):1–9

Printed in the United States
By Bookmasters

Printed in the United States
By Bookmasters